DATE DUE

INDUSTRIAL PRODUCTION MODELS

A THEORETICAL STUDY

BY

SVEN DANØ

PROFESSOR OF MANAGERIAL ECONOMICS
UNIVERSITY OF COPENHAGEN

WITH 71 FIGURES

1966

SPRINGER-VERLAG NEW YORK INC.

© 1966 by Springer-Verlag/Wien

Library of Congress Catalog Card Number: 66-15847

Printed in Austria

Title-No. 9172

In grateful memory of
FREDERIK ZEUTHEN

Preface

This book is a result of many years' interest in the economic theory of production, first aroused by the reading of Professor ERICH SCHNEIDER's classic *Theorie der Produktion*. A grant from the Danish-Norwegian Foundation made it possible for me to spend six months at the Institute of Economics, University of Oslo, where I became acquainted with Professor RAGNAR FRISCH's penetrating pioneer works in this field and where the plan of writing the present book was conceived. Further studies as a Rockefeller fellow at several American universities, especially an eight months' stay at the Harvard Economic Research Project, and a visit to the Unione Industriale di Torino have given valuable impulses. For these generous grants, and for the help and advice given by the various institutions I have visited, I am profoundly grateful.

My sincere thanks are also due to the University of Copenhagen for the exceptionally favourable working conditions which I have enjoyed there, and to the Institute of Economics—especially its director, Professor P. NØRREGAARD RASMUSSEN—for patient and encouraging interest in my work. I also wish to thank the Institute's office staff, Miss G. SUENSON and Mrs. G. STENØR, for their constant helpfulness, and Mrs. E. HAUGEBO for her efficient work in preparing the manuscript, which was completed in the spring of 1965.

Finally, I am indebted to The University of Chicago Press (as publisher of the *Journal of Political Economy*) and the Harvard University Press for permission to quote from various books and articles, and to Professor HOLLIS B. CHENERY, who has kindly allowed me to quote material from his unpublished doctor's thesis.

Copenhagen, April, 1966

SVEN DANØ

Contents

Chapter I

Introduction

1. Entrepreneurial Behaviour and Optimization Models

The central subject of economic science is the allocation of scarce resources. From this point of view the main body of economic theory falls into two parts: the theory of production, which deals with problems of allocation in production, and the theory of consumption or demand.

The economic theory of production, therefore, is as old as economics itself. The modern approach to problems of allocation in production dates back to the Ricardian analysis of income distribution. The subsequent introduction of marginal analysis, a mode of thinking which was embryonically present in RICARDO's distribution model although he did not explicitly apply differential calculus, prepared the way for the neoclassical theory of the firm with its later developments and also made possible the application of such concepts as aggregated production functions, marginal productivities, etc. in theories of income distribution, economic growth, and other fields of macroeconomic analysis.

In microeconomics, the theory of production is often defined in a broad sense as identical with *the theory of the firm*. Its purpose is to describe and explain entrepreneurial *behaviour*, usually in terms of an optimization model where some objective function (e.g. the profit function), which by hypothesis is the criterion of preference underlying the decisions of the firm, is maximized subject to the technological and market restrictions on the firm's behaviour.

For example, the familiar short-run cost function for a single-product firm in a competitive market is derived from the hypothesis that the firm minimizes total cost—i.e., maximizes profit—for each particular level of output with the production function as a side condition; the optimal level of output is then determined by maximizing revenue minus cost. (Alternatively, the solution may be determined in one step without using explicitly the concept of the cost function, maximizing profit subject to the production function). This procedure solves the allocation problem—how much to produce, and by what combination of inputs—for a given set of prices; moreover, by examining the response of the equilibrium position to changes in product price and factor prices respectively, we can trace the firm's supply curve for the product and the demand function for each input.

More generally, the behaviour of a multi-product firm selling its products and buying factors in imperfect markets can be analysed within a model where the side conditions are the production function (or functions), the demand functions for the products, and the factor supply functions (as seen by the producer). The solutions to the system of side conditions—more strictly, the non-negative solutions—represent the range of feasible alternatives open to the firm, i.e., the range of economic choice, and the firm's behaviour is deduced from the hypothesis that, in any situation, the entrepreneur prefers that solution which maximizes the particular objective function (for example, short-run profit), and acts accordingly. The implications of this hypothesis with respect to price and production policy can be tested empirically.

An alternative interpretation of models of this type, more relevant in managerial economics where the emphasis is on optimization in production planning rather than on entrepreneurial behaviour, is that the model is a *normative* one in the sense that it aims at determining the optimum allocation for a given criterion of optimality (objective function), the latter having been selected beforehand on its own merits rather than as a working hypothesis which may or may not turn out to be in accordance with actual behaviour.

2. Input-Output Relationships

In a narrower sense, the theory of production is a *theory of production functions*, concentrating on the technological relations between inputs and outputs in production with special reference to the possibilities of substitution.

While production technology as such does not, strictly speaking, fall within the province of economics, it has an economic aspect in so far as the production function permits of economic choice; to the extent that this is the case, the study of technical input-output relationships becomes a basic concern of the economist's and a prerequisite of dealing with problems of allocation in production, whether for the purpose of analysing entrepreneurial behaviour or with a view to normative optimization models and their practical applications to production planning. In either case the production relations are a fundamental part of the model—from a formal point of view they enter among the side conditions of the optimization problem—so that the shape and structure of the production function has immediate bearing on the allocation problem; in particular, the number of technological equations relative to the number of inputs and outputs is crucial to the range of choice and to the type of substitution that is feasible. The detailed study of production relations, therefore, cannot be dismissed as a matter belonging entirely under the province of engineering, though the economist will naturally have to rely heavily on engineering data for empirical information.

It is the theory of production in this restricted sense that is the main theme of the present study. The purpose of the book is to throw some light on the quantitative relations between the various inputs and outputs of industrial production processes taking place within given plants, where variable factors cooperate with fixed capital equipment.

Problems of optimal allocation will be dealt with, primarily in order to illustrate the range of economic choice permitted by the models; for this reason the examples of optimization will be based upon the simplest possible assumptions, namely, short-run profit maximization under fixed prices of inputs and outputs. This is not to be taken as flat acceptance of a particular hypothesis on entrepreneurial behaviour, nor is it postulated that the assumption of fixed prices is a realistic one. The various optimization models in the book are not at all intended to give realistic descriptions of behaviour but are meant to demonstrate the possibilities of economic choice under certain technological restrictions which the firm will have to respect whatever the type of market and no matter what pattern of behaviour the firm chooses to follow. It is beyond the scope of the present study to indicate what the decisions of the entrepreneur will be on different hypotheses on the firm's objective and under more complicated market conditions. It may be argued in any case that the maximum profit solution to a problem of allocation in production, and particularly the least-cost solution to a factor substitution problem, is always of interest in the more practically oriented field of managerial economics, not only because of its "normative" character—as witness the many practical applications of linear programming, where profit maximization as the criterion of optimality is usually taken for granted—but also as a standard of comparison with solutions derived from alternative objectives.

The analysis being confined to productive processes that take place within given plants, i.e., to short run economic choice in production, new investment in productive equipment will be neglected. It may be objected that, even in the comparatively short run, actual changes in the factor combination are often accompanied by greater or smaller changes in the fixed equipment, that is, by the introduction of new fixed factors[1], and that substitution in this sense is of greater economic consequence than minor adjustments of variable inputs within an unchanged plant. Certainly the concept of a static short-run production function representing a given method of production with given equipment is inadequate as a description of a firm constantly undergoing technological change. However, this is no reason why economic choice in short-run situations with given equipment should be ignored. After all, even a rapidly expanding firm has to make plans for the operation of the plant such as it is at its present stage of development, taking account of the possibilities of input substitution and economic choice in general; moreover, such plans of operation for hypothetical alternatives with respect to productive equipment are essential to rational investment decisions.

Inventory accumulation will also be disregarded: no distinction is made between production and sales, or between quantities of materials purchased and quantities used as input.

---

[1] This problem of short-run alterations in existing plants and their effect on the firm's cost curve is treated by STIGLER (1939), pp. 317 ff.

A microeconomic theory of production models may aim at describing input-output relationships for the production unit as a whole—i.e., the plant—or the primary object of analysis may be the individual technical process. Just as the financial control unit—the firm—may be composed of several plants, the overall operation of the plant can usually be divided up into a number of distinct, well-defined technical processes—associated, for example, with particular units of equipment or with particular outputs—each of which can be described in terms of a particular production model. Much of the literature neglects this complication, more or less tacitly assuming that the product or products are made in one process or, at least, that the integration of the several processes presents no problems as far as the production function and optimal allocation are concerned.

The present study will deal mostly with production models referring to individual technical processes within the plant, although the theoretical analysis will be rounded off with a brief treatment of the integration problem. To keep the analysis at the lowest possible level of aggregation seems likely to be the more fruitful approach. The concept of technological substitution refers logically to the more fundamental unit of the individual process, and so do most of the engineering data from which empirical evidence may be drawn.

Chapter II

Some Fundamental Concepts

1. Production and the Concept of a Process

We shall now proceed to the theoretical analysis of industrial production processes.

Industrial *production* can be defined as the transformation of materials into products by a series of energy applications, each of which effects well-defined changes in the physical or chemical characteristics of the materials. The production of commodities is organized in technical units called *plants*, a plant being a technically coordinated aggregate of fixed equipment under common management.

The total productive activity that takes place within a plant can be broken down into separate *processes*. This can be done more or less arbitrarily in many different ways, depending on the purpose of the analysis and the degree of simplification required, and no clear-cut and universally accepted definition of a productive process exists; it is to a large extent a matter of convenience[1]. If the distinction between plant and process is based on a division of the fixed equipment, as is frequently the case in economic analysis, the concept of a process may refer to the aggregate of all units of a single durable factor used by the plant[2], e. g. machines of a specified type, or to major divisions of the plant such as departments[3], groupings of technologically complementary equipment[4], etc. Further subdivision by products may be called for in cases of joint production where the several products are interrelated only in that they have to share the capacity of the fixed equipment; in such cases the production of each commodity is better regarded as a separate process (or chain of processes). For the purposes of the present study it has been found convenient to use the term "process", as distinct from the total productive activity of the plant, somewhat loosely to mean any specified treatment to which materials are subjected and which brings them further towards the final product or products. A productive

---

[1] For a thorough discussion of the process concept see CHENERY (1953), pp. 299 ff.

[2] This is the concept of the "stage", the basic unit in JANTZEN's theory of cost; cf. BREMS (1952 a, b).

[3] CARLSON (1939), pp. 10 f., illustrates this by the example of a sawmill which, for the purpose of production and cost analysis, is conveniently divided into three technical units corresponding to the processes of power production, sawing, and planing.

[4] Cf. CHENERY (1953), p. 299.

process in this sense will generally be associated with particular items, or aggregates of complementary items, of fixed equipment whose services represent the fixed or scarce inputs of the process[1].

2. The Variables: Inputs and Outputs

(a) In order for a productive process to be fully defined it is necessary to specify very carefully the inputs going into the process and the outputs, or products, resulting from it, as well as to specify the technology of the process itself.

The *outputs*, to be denoted in the following by x_j $(j = 1, 2, \ldots, n)$, are the economic goods produced in the process. Waste products, as long as they are disposed of as worthless, do not count as outputs in an economic sense, although they may come to do so should they become valuable such as to be associated with a positive (market or accounting) price. It should be borne in mind that outputs, as defined here, refer to particular processes rather than to the plant or firm as a whole so that an output is not necessarily a finished marketable product; the output of an intermediate process is an input of the next stage of processing, just as some of the inputs of material may represent intermediate products made in a preceding process rather than primary raw materials.

The *inputs* of a productive process, to be indicated by the symbols v_i $(i = 1, 2, \ldots, m)$, are the measurable quantities of economic goods and services consumed in the process: materials, labour, energy and other current inputs purchased and used by the firm, as well as the services of the fixed equipment, e.g. machine time. Only such factors as are subject to the firm's control are to be included among the inputs; although, for example, climatic conditions may affect the level of output in certain processes so that the climate may be termed a factor of production in a very broad sense, it is not an input either in a technical or in an economic sense.

Each input must be technologically fully specified and quantitatively measurable in well-defined, homogeneous physical units. The familiar broad categories of land, labour, and capital may be useful in a highly idealized model of production in the abstract, if used and interpreted with the utmost care, but they are grossly inadequate for a detailed, empirically meaningful analysis of concrete productive processes on the microeconomic level. Different kinds of labour should be regarded as distinct inputs when they perform different tasks in the process (and, perhaps, are paid differently). Different grades of a raw material should be treated as so many qualitatively distinct inputs (possibly substitutable for one another). For similar reasons, the services of different items of fixed capital equipment must be kept apart in the analysis, except that an aggregate of technologically complementary

---

[1] This concept of a process should not be confused with a method of production characterized by fixed coefficients, known in the linear programming literature as a "(linear) process" or an "activity"; the latter term will be used in Ch. III. A productive process as defined above may be described in terms of an activity if, and only if, all inputs are limitational with constant technical coefficients.

equipment may for convenience be treated as a single factor whose services represent a single input, a procedure applicable to any complex of inputs which for technical reasons will have to vary proportionately. In any case the units in which an input is measured will be required to be physically homogeneous such as to exclude qualitative changes from the analysis; only quantitative variations in the number of identical units of each single input will be considered[1].

Current inputs will be measured in the units in which they are purchased by the firm, i.e., in such units as are usually associated with market prices. For example, fuel should be measured in tons of coal, gallons of oil, etc., not in terms of heating value (calories). In the case of labour, man-hours will be used as unit of measurement whether the workers are paid by the hour or per unit of output produced (piece-work rates)[2]; in the latter case a given piece-work rate merely implies that the hourly wage is no longer constant but varies with the level of output and the factor combination (unless the labour coefficient is a technologically given constant)[3].

As to capital inputs, the various items of capital equipment *as such* do not appear as input variables in models of production processes taking place

---

[1] As an example of a production model which does not fulfil this requirement we may quote from STIGLER (1939), p. 307: "The law of diminishing returns requires full adaptability of the form, but not the quantity, of the "fixed" productive services to the varying quantity of the other productive service. To use a well-known example, when the ditch-digging crew is increased from ten to eleven, the ten previous shovels must be metamorphosed into eleven smaller or less durable shovels equal in value to the former ten, if the true marginal product of eleven laborers is to be discovered." The fixed factor, thus defined, is not a well-defined input in a technical sense at all; any variation in the level of output is accompanied by qualitative changes in one of the inputs, disguised as quantitative changes by the use of capital value as the unit of measurement. Clearly this procedure is quite unsuitable for analysing production processes within a given plant where variable inputs cooperate with a given stock of specified fixed equipment.

[2] To measure the input of labour in terms of output produced because this is the basis of piece-work rates would be to dodge the problem of defining and determining the quantitative relationships between inputs and outputs. Carried to its logical conclusion the procedure leads to a production relation of the form

$$x = v,$$

which is formally a case of limitationality (the coefficient of production being $= 1$) but which is totally devoid of empirical content. The quantity of an input should always be measured independently without reference to the resulting output and the quantities of other inputs employed—a rule not always observed in, for example, the theory of distribution where working effort, ultimately defined and measured in terms of productivity, is sometimes counted as a separate dimension of the supply of labour.

[3] Let x be output, v input of labour in man-hours, and w the (given) piece-work rate. Then the hourly wage will be

$$q = \frac{w \cdot x}{v}$$

where w is constant whereas x is a function of v and—in the case of factor substitutability—of other inputs as well.

Moreover, the piece-work rate will appear as a parameter in the production function (the higher the rate, the higher output per hour). We shall not, however, consider such variations in the remuneration of labour.

within a given plant; being given in number, they are fixed factors. In so far as they are productive only in an indirect manner, i.e., by their mere presence, they are represented by parameters in the production model or by the shape of the production function. Buildings are a case in point. On the other hand, the *services* of such capital equipment as is directly productive in the sense that the rate of output varies with the utilization rate should be explicitly represented in the model by a specific input variable—a point not always observed in the literature. For example, even if the stock of machines of a certain kind is fixed, the flow of services which the machines yield during the period considered—as measured in machine hours—must be specified in the production function as a input variable along with the current inputs because it affects output per period (i.e., per unit of calendar time) in a similar way. More specifically, the input variable should represent the number of machine hours actually *used* during the period since idle time does not contribute to output. Even when the fixed factor itself is physically indivisible (e.g. one machine) its services are usually divisible in the time dimension so that it is necessary to distinguish between the amount of services available—i.e., the capacity of the factor—and the amount which is used during the period; the latter is the input to be specified as a variable whereas the former represents an upper limit to it. Thus the services of the fixed factor represent an input which it not fixed in the sense of being constant but which may, more appropriately, be termed a scarce input because of the capacity limitation. It is only in the highly special case of the services being completely indivisible that they need not be specified explicitly as an input variable but may be taken as implicit in the shape of the production relations.

Since the costs of the fixed factors as such are independent of the degree of utilization, being in the nature of "historical" costs derived from the prices at which the equipment was bought, no market prices are associated with their services, which can therefore, in a short-run model, be regarded as free inputs up to capacity. It is possible to determine imputed prices in the sense of marginal opportunity costs associated with the services of fixed factors ("shadow prices"), but they are not market prices and should not appear in the cost function; nor are they in any way related to the fixed costs.

(b) Production being a time-consuming process, a complete technical description of a productive process would have to include an account of the configuration of inputs and outputs in time. However, the time lags between these variables—i.e., the length of the production period—can usually be neglected in an economic study of short-run production functions, especially when the productive processes analysed are continuous and repetitive with a constant instantaneous time rate of output. The present analysis will be confined to timeless production in the sense that inputs and outputs are referred to the same period. In a short-run analysis we also ignore the links between consecutive production periods due to the fact that some factors of production (the fixed plant) are durable. Nor will we deal with the coordination of operations in time, i.e., the proper timing of arrivals of units requiring processing and the order in which the various jobs of processing are to be performed[1]; for one thing these problems do not arise when only a single process or operation is considered.

There is one aspect of the time factor, however, which will have to be taken into account even in short-run static analysis. Inputs and outputs, and thus also cost and profit, are time rates, having the dimension of physical units *per period of time*. As BREMS has put it, "That cost is a time rate is ultimately apparent from what economists have to say about cost. If cost were measured in dollars, and if quantity were measured in pounds, yards, or another physical unit, one could always increase quantity with a proportional increase in cost—simply by letting a longer time elapse! Obviously this is not what economists mean when they talk about the cost-quantity relationship. What they mean is this: How does an increase in quantity produced (or sold) within a unit period of time affect costs within that period?"[2]

The quantities of inputs and outputs can be varied in time as well as in space—for example, total labour input in man-hours is number of workers times number of hours worked—and in order to specify the set of feasible combinations, i.e., the production function, we must first specify the kind of variation that we wish to consider. This we do by measuring inputs and outputs in physical units per period of given (though arbitrary) length—for example, per week, month, or other unit of calendar time—thus confining the analysis to "spatial" variations and such variations in the time dimension as are feasible within the given period. The latter kind of variation consists in varying the utilization rate, the proportion of "active" hours to total available hours per period of calendar time. In choosing the length of period we also fix the capacity of the plant or fixed equipment, capacity being dependent on the number of plant or machine hours available per period. Had the variables of the production model been measured in absolute physical units, there would be no capacity limitation since output could always be expanded indefinitely in the time dimension (presumably at constant returns to scale), which is clearly not the kind of variation we are interested in analysing, although we must allow for the possibility of less than full-time utilization of capacity.

In economic literature—particularly in examples of the law of variable proportions in agriculture—inputs and output are sometimes defined in terms of absolute physical units (e.g. corn yield in bushels, fertilizer in pounds); in such cases, however, it is implicitly understood that the quantities refer to a given period (a year). The fact that an identical repetition of the process—applying twice the amounts of labour, fertilizer, etc. to two years' services of the same acreage of land—will double the yield of corn is hardly an interesting one; what matters is the relationship of inputs and output within a period of given length.

---

[1] Cf. FRISCH (1956), p. 25. For treatment of these problems of "scheduling" or "sequencing" the reader is referred to the literature on operations research; cf., for example, CHURCHMAN, ACKOFF and ARNOFF (1957), Ch. 16.

[2] Cf. BREMS (1951), p. 55.

3. The Technological Relations: The Production Function

For economic analysis, a well-defined productive process with specified inputs can be described in more or less idealized form in terms of a *production function* or, more generally, a *production model*, i.e., a system of quantitative relationships expressing the restrictions which the technology of the process imposes on the simultaneous variations in the quantities of inputs and outputs. To ensure uniqueness it is assumed that, for each particular factor combination, the inputs are organized in a given specified manner as prescribed by the technology; indeed, this—together with the requirement that the list of specified inputs (and outputs) shall remain the same over the range of variation considered—may be taken as a definition of a given constant technology[1].

A priori there is no reason why the production model should always consist of one equation only. Single-equation models are an important class of special cases; for example, the case of a single product with all inputs continuously substitutable is represented by a single production function expressing x as depending upon the v's. But this will not do in cases where some of the v's are related by technological side conditions so that there are fewer than m independent input variables, or where the quantities consumed of certain inputs are uniquely related to the amount of output produced. It is sufficient here to mention fixed coefficients of production (limitational inputs) as an example. In the more general case, then, the production model for a single-output process can be written

$$x = f_k(v_1, v_2, \ldots, v_m) \qquad (k = 1, 2, \ldots, M) \qquad (1)$$

where $1 \leq M \leq m$. More generally still, allowing for joint production, the model has the form

$$F_k(x_1, x_2, \ldots, x_n; v_1, v_2, \ldots, v_m) = 0 \qquad (k = 1, 2, \ldots, M) \qquad (2)$$

where the number of equations cannot exceed $m + n - 1$ because the system must obviously have at least one degree of freedom such as not to be completely determined from the outset. The production model is often referred to as the production function even when it consists of more than one input-output relation. From an economic point of view the number of equations as compared with the number of variables—i.e., the number of degrees of freedom which the technology permits—is one of the most important characteristics of the technology of a particular process because it represents the dimension of the range of economic choice[2].

This is not brought out explicitly if the input variables of the model are defined as *available* amounts of the respective inputs, $w_1, w_2, \ldots, w_m$, rather

---

[1] Cf. FRISCH (1956), pp. 20 ff. Conversely, FRISCH defines technological change as irreversible change in the organization of a given set of factors (and thus in the shape of the production function) or qualitative change in one or more factors (new inputs replacing others in the list of specified inputs).

The existence of particular factor combinations where one or more inputs is zero (i.e., not used) is not ruled out by the definition as long as the criterion of reversibility is observed.

[2] This also holds for linear production models with discontinuous substitution, where the activity levels appear as variables in addition to the v's and x's.

than the amounts actually *consumed* in the process, $v_1, v_2, \ldots, v_m$. Since, by definition, there are no technological restrictions on the variation of the quantities that are available, the w's can be treated as independent variables. It follows that a single-product model can always be written in the form

$$x = \varphi(w_1, w_2, \ldots, w_m) \tag{3}$$

where, for uniqueness, x is to be interpreted as the *maximum* amount of output that can be produced with any given set of input amounts available[1]. In this manner, the number of equations in the model can be reduced to one regardless of the particular technology that characterizes the process.

However, quite apart from the fact that the procedure is not immediately applicable to cases of joint production[2], the simplification thus obtained is more apparent than real. The analytical advantage of writing the model in the form (3) is somewhat limited since the function does not possess continuous derivatives except when all inputs are continuously substitutable for one another, in which case the model consists of a single equation anyhow. Moreover, the formulation (3) tends to blur the fundamental issue of substitutability. In the case of fixed coefficients of production, for example, the inputs may be available in any amounts but they will be used in fixed proportions; the maximum level of output is determined by the scarcest input, relative scarcity being defined with reference to the production coefficients. The basic characteristic of the production technology is the absence of factor substitution, and this is directly brought out only by a model of type (1) expressing proportionality between output and the amount of each particular input consumed. The production function (3) must be thought of as derived from these more fundamental technological relationships. Besides, for cost accounting as well as for the analysis of the firm's demand for factors of production, the relevant variables are the input amounts actually used rather than the quantities that happen to be available[3].

For these reasons the analysis of production processes is in general best carried out in terms of models of type (1) or (2), using $v_1, v_2, \ldots, v_m$ for input variables rather than $w_1, w_2, \ldots, w_m$. The latter are relevant only in the case of the fixed factors, whose services are not bought in the market but available to the firm in given amounts per period. These limiting amounts should appear in the model only as upper bounds to the amounts of services actually used. This means that, in addition to the technological relations, we have explicit *capacity restrictions* in the form of inequalities

$$v_s \leq \bar{v}_s \tag{4}$$

---

[1] Cf. FRISCH (1953), pp. 1 ff. The production function is defined in this way by SAMUELSON (1947), Ch. IV, and by many other authors.

[2] For one thing, $x_2, x_3, \ldots, x_n$ cannot always be given arbitrary values (with a view to maximizing x_1) when the values of the independent input variables $w_1, w_2, \ldots, w_m$ have been picked. In the case of joint production with fixed output proportions and fixed input coefficients there will be an upper limit to the amount of each particular product that can be produced with given input amounts available.

[3] Cf. the concept of real cost, from which the cost function is derived by multiplying by the factor prices and adding.

where v_s is an input variable in the production function—representing, for example, the number of machine hours utilized per period of calendar time—whereas $\bar{v}_s$ represents available machine hours per period $(=w_s)$[1]. The two coincide only when the services are completely indivisible so that we have $v_s = w_s = \bar{v}_s$, in which case the fixed factor is represented in the production function not by a variable v_s but by the parameter $\bar{v}_s$ [2].

The production model is defined over the non-negative region since negative values of x_j or v_i are devoid of economic meaning. This means that the variations of the input and output variables are also constrained by the *non-negativity requirements*

$$v_i \geq 0, \quad x_j \geq 0 \tag{5}$$

although these restrictions are not always written out explicitly.

4. Efficient Production and Economic Optimization

With maximum short-run profit as the criterion of optimal allocation, the optimization problem with which the firm is confronted is that of maximizing the profit function

$$z = px - \sum_{i=1}^{m} q_i v_i$$

—or, in the multi-product case,

$$z = \sum_{j=1}^{n} p_j x_j - \sum_{i=1}^{m} q_i v_i$$

—subject to the technological restrictions (1) or (2) (the production model), to capacity limitations (4), and to the non-negativity requirements (5)[3]. It is assumed that the product prices p_j and factor prices q_i are constant (in the case of fixed factors, usually equal to zero) so that the profit function is linear[4]. The side conditions constitute a system of equations and in-

---

[1] *Capacity restrictions* on *inputs* representing the services of the fixed factors, $v_s \leq \bar{v}_s$, lead to an upper limit to the amount of *output* that can be produced per period, $x \leq \bar{x}$, where $\bar{x}$ is called the (maximum) *capacity* of the plant (or process).

[2] Buildings may also set capacity limits although their services are not specified as input variables (cf. above); there is an upper bound to the number of machines and other equipment that can be housed in a factory building and thus to the amount of output that can be produced. However, in a model which assumes that the number of machines and other items of capital equipment is fixed, limitations of this kind never become effective and should be left out as redundant.

[3] The profit function as defined here refers to a single process rather than to the plant as a whole. If the operation of the plant is described in terms of several interrelated processes, a separate optimization model may be formulated for each process, using accounting rather than market prices for intermediate goods. The question whether suboptimization of this kind leads to maximum aggregate profit is discussed in Ch. IX.

[4] For greater generality the constant prices would have to be replaced by demand functions relating the p_j to the x_j and by factor supply functions $q_i = q_i(v_i)$. Selling effort as a parameter of action may be introduced by letting the cost of advertising appear as a parameter in each demand function and also as an additional term in the cost function. These complications will be neglected because we are primarily interested in the technical production relations as such. The problem of product quality, which raises further complications, is dealt with in Ch. VIII.

equalities defining the set of feasible points; all points—i.e., quantitative combinations of inputs and outputs—which satisfy the system represent alternative feasible solutions to the optimization problem and that point among them which yields the greatest profit is the optimal solution[1].

The solution to this problem represents both the optimal (least-cost) allocation of inputs and the optimal level of output (in the multi-product case, the optimal product mix). Instead of solving these two allocation problems simultaneously, they may be treated separately in two steps. The first is to determine the cost function by minimizing total (variable) cost of production

$$c = \sum_{i=1}^{m} q_i v_i$$

subject to (1) or (2), (4), and (5) for given parametric x or x_j; the solution gives the locus of least-cost points in input space, i.e., the optimal values of the v_i in terms of the x_j. Inserting these values in the linear cost expression we get c as a function of the output parameters,

$$c = c(x) \quad \text{or} \quad c = c(x_1, x_2, \ldots, x_n),$$

which is the familiar cost function of economic theory.

The second step is to maximize the profit expression

$$z = px - c(x) \quad \text{or} \quad z = \sum_{j=1}^{n} p_j x_j - c(x_1, x_2, \ldots, x_n)$$

such as to determine the level of output (in the case of joint production, the optimal product mix).

It may be argued that this kind of two-step analysis, frequently used in textbooks as a mere expository device, can be dismissed as representing a roundabout way of getting to the optimum solution of the overall optimization problem; from an analytical point of view the cost function as an intermediate stage can be dispensed with altogether when the purpose of the analysis is to find the point of maximum profit. The tendency in recent management science and operations research is towards taking this view, the more so because the cost curves to be derived from linear programming and related models of production are analytically awkward because of discontinuities. On the other hand, the separation of the problem into two steps may be defended on the ground that it reflects a similar separation of management functions, since in practice the decisions with respect to the level of output and the adjustment of the factor combination are often made separately by different administrative units within the firm[2][3], although the

---

[1] The mathematical methods of maximizing some function subject to side conditions vary with the formal nature of the problem (the types of functions and constraints involved). A brief account of the various methods of constrained optimization—the Lagrange-multiplier device, linear programming, etc.—is given in Appendix 1.

[2] Cf. SHEPHARD (1953), p. 9.

[3] It may also be argued that the degree of realism in the underlying assumptions (profit maximization and constant prices) is different between the two kinds of decisions: even when

two sets of production decisions must of course be coordinated in some way to ensure the best results.

Obviously, as mentioned above, the range of choice open to the firm in optimizing production depends on the number of degrees of freedom in the production model (1) or (2) [1]. In the case of only one degree of freedom everything is determined up to a scale factor representing the level of operations. The optimization problem is economically interesting only when the production function has enough degrees of freedom to allow substitution, i.e., when the same level of output or the same batch of outputs can be produced by alternative combinations of inputs[2] (factor substitution) or when a given input combination can produce alternative combinations of outputs (product substitution).

According to the conventional division of labour between the technician and the economist, it is the former's job to provide the production relationships, whereas the latter is responsible for indicating the optimal allocation of resources, in the present case that combination of inputs and outputs which will maximize profit subject to (1) or (2), (4), and (5). The position of the economic optimum will depend on the product and factor prices.

It may happen, however, that the region of feasible solutions contains points which can be dismissed beforehand regardless of economic considerations—i.e., no matter what the prices are—because they are technologically *inefficient* in the sense that it is possible to produce more of one output without having to produce less of any other output and without using more of any input, or that it is possible to produce the same amounts of all outputs with less of one input and not more of any other input. More precisely, a point $(x_j{}^0, v_i{}^0)$ satisfying (2), (4), and (5) is inefficient if there exists at least one other feasible point $(x_j{}^1, v_i{}^1)$ such that

$$x_j{}^1 \geq x_j{}^0, \quad v_i{}^1 \leq v_i{}^0 \qquad (j = 1, 2, \ldots, n; \; i = 1, 2, \ldots, m) \qquad (6)$$

(not all equalities or the two points would coincide). Clearly an inefficient point cannot represent maximum profit (or minimum cost) since it follows from (6) that

$$\sum_{j=1}^{n} p_j x_j{}^0 - \sum_{i=1}^{m} q_i v_i{}^0 < \sum_{j=1}^{n} p_j x_j{}^1 - \sum_{i=1}^{m} q_i v_i{}^1 \qquad (7)$$

for any set of non-negative prices (not all equal to zero). Conversely, if there exists some point $(x_j{}^1, v_i{}^1)$ for which (7) holds for any set of prices, the given point $(x_j{}^0, v_i{}^0)$ will be technically inefficient as we have

---

the firm does not aim at maximum profit, there is no particular reason why it should not try to manufacture its products in the cheapest possible way, and the factor prices are less likely to vary with the firm's production decisions than the prices of the products.

[1] The inequalities (4) and (5), being in the nature of mere boundary conditions, impose bounds on the values of the variables but do not affect the number of degrees of freedom, i.e., the dimension of the set of feasible solutions to the optimization problem.

[2] That is, by alternative combinations of the *same* inputs—including, of course, combinations in which some of the input variables are zero.

$$\sum_{j=1}^{n} p_j(x_j{}^0 - x_j{}^1) + \sum_{i=1}^{m} q_i(v_i{}^1 - v_i{}^0) < 0$$

which, for $p_j \geq 0$ and $q_i \geq 0$, is readily seen to imply (6). Thus the economically relevant range of economic choice is restricted to the set of technologically *efficient* points, i. e., feasible points which are not inefficient as defined above. For any two efficient points $(x_j{}^1, v_i{}^1)$ and $(x_j{}^2, v_i{}^2)$ we have

$$\sum_{j=1}^{n} p_j x_j{}^1 - \sum_{i=1}^{m} q_i v_i{}^1 \gtreqless \sum_{j=1}^{n} p_j x_j{}^2 - \sum_{i=1}^{m} q_i v_i{}^2,$$

their relative profitability—and thus the choice between them—depending on what the prices are. This is where the economist comes in[1].

Inefficiency in production means that resources are employed in a wasteful manner. Some cases of obvious waste have been ruled out beforehand by our definition of the input variables in the production function as the quantities of inputs actually used in the process: for example, in the case of fixed coefficients of production, combinations where the inputs are in "wrong" proportions so that some of them are not used up do not belong to the production function at all. There are other examples of waste, however, which are compatible with the production function as defined here and which are to be described as inefficient production. One example is a factor combination in which the marginal productivity of some input is negative; such a point is inefficient in that it is possible to produce more by using less of the input in question without using more of other inputs[2].

---

[1] In terms of vector analysis, the efficiency problem is that of *ordering* all vectors

$$(x_1, x_2, \ldots, x_n, -v_1, -v_2, \ldots, -v_m)$$

representing points satisfying (2), (4), and (5) (except that inputs have been given a negative sign). One such vector is greater than another if the inequality $\geq$ holds for each pair of corresponding elements, i. e., if relations such as (6) can be established between them. If this can be done between any two of the vectors, a complete ordering can be established and the optimization problem solves itself without resort to economics as there is only one efficient point. This is unlikely ever to be the case in reality. The vectors must be expected to be only partially orderable, i. e., there is a choice between several efficient points (in some cases, an infinity of such points) and the choice will have to be made on the basis of economic considerations; see CHARNES and COOPER (1961), Vol. I, pp. 294—296. Cf. the analogous concept of PARETO optimality in the theory of economic welfare.

[2] This was of course recognized long before the efficiency concept had been developed by KOOPMANS (1951) in connexion with the linear activity analysis model. See, for example, GLOERFELT-TARP (1937).

Chapter III

Linear Production Models
and Discontinuous Factor Substitution

A. Limitational Inputs and Fixed Coefficients of Production

(a) We shall now deal with the more important types of production functions for processes which produce one single commodity, with particular emphasis on the formal aspects that are connected with factor substitutability.

The simplest kind of production model is that in which all inputs are *limitational* in the sense that no factor substitution is possible; a given quantity of output can be produced by one and only one combination of inputs. This means that the amount of each input used in the process depends on the level of output only[1],

$$v_i = v_i(x) \qquad (i = 1, 2, \ldots, m) \qquad (1)$$

or, written in terms of the inverse functions,

$$x = f_1(v_1) = \ldots = f_m(v_m).$$

This case is also known as *perfect complementarity*: not only does it take some of each input to get a positive product, but the quantities required are uniquely related to output.

An important special case is that of proportionality between x and the v_i,

$$v_i = a_i \cdot x \qquad (1\,a)$$

where the a_i are constant coefficients of production[2] so that the model is characterized by constant returns to scale.

---

[1] The term "limitational" in this sense was first coined by FRISCH (1932). A wider concept of limitationality was later introduced by FRISCH (1953) pp. 3 ff., to include other cases of multi-equation models; cf. Ch. V.

[2] The case of fixed coefficients of production is usually associated with the name of WALRAS. In the first editions of the *Éléments*, the coefficients of production in his model of general equilibrium were technologically given constants, an assumption made primarily for the sake of simplicity. In the definitive edition, however, the assumption of fixed coefficients in Lesson 20 was a purely provisional one, to be dropped later in Lesson 36 where the marginal productivity theory was introduced explicitly. Cf. WALRAS (English ed., 1954), pp. 239 f., 382 ff., and 549 ff. Constant technical coefficients—though at the industry level rather than that of the individual firm or process—are also characteristic of LEONTIEF's input-output models, cf. LEONTIEF (1941, 1953).

From an economic point of view this type of production model is not particularly interesting because the corresponding optimization problems are trivial. There is no economic problem of cost minimization for given level of output since the factor combination is technically determined. The production relations (1) define a one-dimensional curve in m-dimensional input space—in the special case (1 a), a straight line—each point of the curve being the isoquant for a particular value of x; and the expansion path, i. e., the locus of optimal (least-cost) factor combinations, will coincide with this curve no matter what the factor prices are.

The only optimization problem which involves economic considerations is that of determining the optimum level of output. In the linear limitational case (1 a) the total variable cost of production

$$c = \sum_{i=1}^{m} q_i v_i = \left(\sum_{i=1}^{m} q_i a_i \right) \cdot x,$$

The standard textbook example of production processes with limitational inputs is *chemical* reactions. It is widely held by economists that the technology of a chemical reaction can always be described by a set of constant input coefficients determined by stoichiometric proportions, an idea which appears to date back to PARETO, whose somewhat irrelevant concrete example— water made from hydrogen and oxygen—has shown a curious persistence in economic literature. (Cf. PARETO (1897), p. 102, and (1927), pp. 326 f.) This would indeed be true if chemical reactions were always characterized by instantaneous 100% conversion of materials fed into the process in stoichiometric proportions. Actually, however, it may not be economical to allow the process to go to completion because the rate of conversion is a decreasing function of reaction time (although in some cases the use of catalysts will speed up the process); moreover, the rate will depend on process conditions such as temperature, pressure, initial concentrations, etc. even if the process equipment is fixed. In the second place, even if sufficient time is allowed for the process, the chemical equilibrium which is eventually attained is seldom characterized by 100% conversion of the materials; in a reversible process the degree of conversion—i.e., the yield of the process—is subject to variation by manipulating the process conditions. Thirdly, the process may be disturbed by side reactions between the components. Some of the process conditions through which the yield can be controlled are directly related to the inputs of materials (viz. the initial concentrations), others (e.g. temperature) are connected with the input of energy. All this means that optimum operation is not uniquely determined but will depend on the prices.

Still, concrete examples of chemical processes exist in which the input coefficients are stoichiometrically determined or at least approximately fixed for a given plant. Numerical data for a plant producing a variety of chemicals (ammonia, nitric acid, ammonium carbonate, polyvinyl chloride, etc.) are reported by BARGONI, GIARDINA and RICOSSA (1954). These compounds are made from wholly or in part the same chemicals (acetylene, carbon dioxide, hydrogen, chlorine, etc.) which combine in fixed proportions for each product.

Mechanical processes where the product is manufactured on constant-speed machines attended by a constant number of operators and where machine time and raw-material input are constant per unit of output can be described by a model of the type (1 a). (If alternative machinery is available the case will be characterized by discontinuous substitution (Ch. III, B) whereas variable machine speed leads to continuous factor substitution, cf. Ch. VII, B.) For an empirical example see FRENCKNER (1957), pp. 81 ff.: a machine-tool factory turns out three different products (machines) each of which requires constant amounts of inputs per unit of output, including constant machine times on the fixed equipment (lathes, milling and grinding machines, etc.). Similarly, the extensive study of the cotton textile industry due to ANNE P. GROSSE (1953) points to the conclusion—supported by a wealth of engineering information—that cotton textile manufacturing is characterized by the absence of factor substitution at all stages (*op. cit.*, pp. 369 f.).

as well as total revenue, $p \cdot x$, is a linear function of x (all prices assumed to be constant). Assuming that p is greater than unit cost, this means that profit could be expanded indefinitely but for the existence of fixed factors whose capacities set upper limits to the level of output that can be produced. Let v_1, for example, represent the services of a fixed factor available in the amount $\bar{v}_1$ per period. The optimal level of output is found by maximizing gross profit (revenue minus variable cost, q_1 being $=0$)

$$z = px - \sum_{i=2}^{m} q_i v_i$$

subject to the production relations

$$v_i = a_i x \qquad (i = 1, 2, \ldots, m)$$

and to the capacity limitation

$$v_1 \le \bar{v}_1,$$

i.e., by maximizing the linear function

$$z = \left(p - \sum_{i=2}^{m} q_i a_i \right) \cdot x$$

subject to the linear inequality

$$a_1 x \le \bar{v}_1 .\,^1$$

The optimal solution obviously is

$$x = \bar{v}_1 / a_1,$$

as illustrated in *Fig. 1a–b* [2]. If the second factor had also been fixed, with

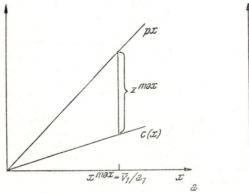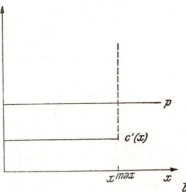

Fig. 1a–b

---

[1] JANTZEN's "Law of Capacity"—cf. JANTZEN (1924), pp. 5 ff., or (1939), pp. 5 ff., and BREMS (1952a, b)—is based on a production model of this kind. JANTZEN's cost function includes the fixed cost per period—defined as the price of the fixed equipment divided by its useful life as measured in production periods—and the law specifically refers to the shape of the corresponding unit cost curve. Cost per unit of output is equal to operating unit cost, which is constant (cf. above), plus fixed cost per unit of output, which varies inversely with x.

[2] This is an example of a "corner" maximum, not—like the "tangency" solutions of maximization problems with side conditions in the form of equations—characterized by a marginal equality; as shown in Fig. 1b, marginal revenue p is greater than marginal cost: $p > c' = \Sigma q_i a_i$.

capacity $\bar{v}_2$, the solution would have been either $x = \bar{v}_1/a_1$ or $x = \bar{v}_2/a_2$, the lesser of these upper bounds being the effective capacity limit.

(b) In determining the optimum level of output, the services of the fixed factors are treated as free inputs ($q_1 = q_2 = 0$) because their contribution to the cost of production is in the nature of a fixed charge associated with their presence (i.e., with the capacities $\bar{v}_1$ and $\bar{v}_2$) independent of the degree of utilization (v_1 and v_2). However, since the profit that can be obtained depends on the capacity limits (or on the degree of utilization of given capacities), the services of a fixed factor may be valued on an *opportunity cost* basis. Such a valuation is provided by the *"shadow prices"* which are implicit in the model.

In the example above there were two fixed factors, $v_1 \leq \bar{v}_1$ and $v_2 \leq \bar{v}_2$. Let the first of these limits be the effective one so that $\bar{v}_1$ represents the bottleneck. Then the optimal level of output is determined by $x = \bar{v}_1/a_1$. Treating $\bar{v}_1$ as a parameter, we have for a marginal change in $\bar{v}_1$ at this point

$$\frac{dz}{d\bar{v}_1} = \frac{dz}{dx} \cdot \frac{dx}{d\bar{v}_1} = \left(p - \sum_{i=3}^{m} q_i a_i \right) \cdot \frac{1}{a_1}$$

which is the shadow price (y_1) of the services of the first factor, to be interpreted as the marginal increase in profit for a unit increase in the capacity $\bar{v}_1$—i.e., the maximum amount the firm would pay for an additional unit of the factor's services—or the opportunity cost of not using the last unit, in terms of forgone profit. The shadow price y_1 is positive unless $p < \Sigma q_i a_i$, in which case production would not be profitable at all.

For $\bar{v}_2$ we have at the optimum point

$$y_2 = \frac{dz}{d\bar{v}_2} = \frac{dz}{dx} \cdot \frac{dx}{d\bar{v}_2} = 0$$

since $dx/d\bar{v}_2 = 0$ at a point where $v_2 < \bar{v}_2$; the marginal value of the services of a fixed factor which is not fully utilized is zero, as we would expect.

The shadow prices represent imputed values rather than market prices; they should not be confused with the historically determined fixed costs associated with the capacity factors, or with the market prices per unit of fixed equipment[1]. However, if the shadow prices are known, they may be applied as accounting prices to determine the optimum level of output by the condition that "net" profit—defined as gross profit minus imputed cost of the fixed factors—shall be zero. We have

$$px - c(x) - y_1\bar{v}_1 - y_2\bar{v}_2 = \left(p - \sum_{i=3}^{m} q_i a_i \right) \cdot \left(x - \frac{\bar{v}_1}{a_1} \right) \leq 0 \text{ for } x \leq \frac{\bar{v}_1}{a_1},$$

that is, "net" profit in this sense is negative except for $x = \bar{v}_1/a_1$ where it is

---

[1] There is a connexion between them, however, in that marginal investment decisions will depend on the current price of new equipment as compared with the firm's estimate of the internal value of the services which another item of equipment will yield over its expected lifetime. The internal valuation will be based upon marginal profit opportunities, i.e., shadow pricing.

zero so that the selling price just covers total "cost"[1]. In this way the optimum point can be determined by a kind of full-cost procedure—only it is the quantity of output that is unknown, the price being given—as shown in *Fig. 2*.

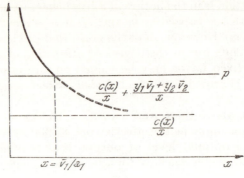

Fig. 2

Alternatively, applying the shadow prices to the amounts of services actually used, v_1 and v_2, rather than to the capacities, we find that total cost in this sense is equal to total revenue at all levels of output,

$$c(x) + y_1 v_1 + y_2 v_2 = \sum_{i=3}^{m} q_i a_i x + \left(p - \sum_{i=3}^{m} q_i a_i \right) \cdot \frac{v_1}{a_1} = p \cdot x$$

so that unit "profit" is zero along the expansion path[2].

---

[1] At the optimum point, then, the total imputed value of available plant services (per period) just absorbs the gross profit margin; cf. the Marshallian concept of quasi-rent.

[2] This result is rather uninteresting except in relation to linear models with discontinuous substitution (several "activities", i.e., alternative sets of fixed coefficients of production, cf. III, B below), where the relative profitability of the various activities can be calculated in this manner; those activities which enter into the optimal solution are characterized by zero profit as defined here, the others by negative profit per unit of output.

Treating the present model as a special case of such a linear programming model, the solution to our optimization problem

$$a_1 x \leq \bar{v}_1$$
$$a_2 x \leq \bar{v}_2$$
$$z = bx = \text{maximum} \left(\text{where } b = p - \sum_{i=3}^{m} q_i a_i \right)$$

can be determined by the simplex method, which also gives the shadow prices as a by-product (cf. Appendix 1). Thus the shadow prices would appear to be of no use since, in determining them, we have also found the optimal level of output. However, the shadow prices might have been determined independently, without recourse to the problem above, by solving the "dual" problem

$$a_1 y_1 + a_2 y_2 \geq b$$
$$f = \bar{v}_1 y_1 + \bar{v}_2 y_2 = \text{minimum},$$

i.e., minimizing the total imputed value of the two capacities subject to the condition that the imputed cost of the fixed factors' services shall not fall short of the gross profit margin. The optimal solution is $y_1 = b/a_1$, $y_2 = 0$. Using these prices we get

$$a_1 y_1 + a_2 y_2 = b,$$

(c) In the literature, the case of production with limitational inputs is more often than not illustrated by L-shaped isoquants for the two-factor case, such as $ABCDE$ in *Fig. 3.*[1]

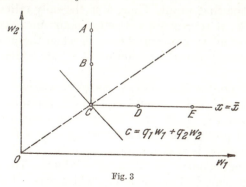

Fig. 3

This is because the input variables of the model are defined as quantities available (w_i), not as the amounts consumed in the process (which would correspond to point C only). Instead of (1a) we then have

$$(v_i =)\, a_i x \leq w_i \qquad (i = 1, 2, \ldots, m) \tag{1b}$$

where the w_i—unlike the v_i—are independent variables. The inequalities (1b) do not constitute a production function $x = \varphi(w_1, w_2, \ldots, w_m)$ determining x uniquely as a function of the w_i; they merely set upper limits to x for given w_i, and any level of output that respects all of these limits is possible. For example, if point B represents the amounts that are available, any level of output requiring the two inputs in a combination on OC is possible. A single-valued production function is obtained by defining x as the maximum amount of output that can be produced with given w_i, i.e.,

$$x = \text{Max } (v_1/a_1) \text{ for } \frac{v_1}{a_1} = \frac{v_i}{a_i} \text{ and } v_i \leq w_i \qquad (i = 1, 2, \ldots, m);$$

since the lowest of the upper limits set by (1b) will be the effective limit to output—hence the name "limitational" inputs—the function φ will have the form[2]

$$x = \text{Min } (w_i/a_i) \qquad (i = 1, 2, \ldots, m). \tag{1c}$$

This is a derived model since the technical coefficients a_i, i.e., model (1a), must be known before (1c) is fully specified. Any point on $ABCDE$ will produce the same quantity of output $\bar{x}$, using the two inputs in the amounts

which is our alternative criterion of profitability (unit "profit" equal to zero). Output should be expanded up to the point where

$$b - \frac{y_1 \bar{v}_1 + y_2 \bar{v}_2}{x} = 0$$

which is satisfied for $x = \bar{v}_1/a_1$.

The mathematical equivalence of the original problem and its dual follows from the duality theorem of linear programming; cf. Appendix 1 and III, B below.

[1] Cf., for example, CARLSON (1939), p. 24, and SCHNEIDER (1958), p. 166.
[2] FRISCH (1953), p. 2.

$v_1 = a_1 \bar{x}$ and $v_2 = a_2 \bar{x}$. At point C the inputs are available in precisely those amounts; at B there is "too much" of w_2 available (w_1 being the minimum factor) so that the excess amount BC will go to waste, whereas point D implies waste of the first input. Thus, C is the only efficient point on the isoquant $ABCDE$: it represents not only maximum output for given inputs—this is true of every point on $ABCDE$ when the production function is defined by (1 c)—but also the minimum of inputs which will produce the given x [1].

In this way the production function $x = \varphi(w_1, w_2)$ can be illustrated geometrically by a family of L-shaped isoquants in the factor plane. Alternatively, as shown in *Fig. 4*, the two-factor case may be illustrated by a family of (kinked) production curves showing x as a function of, say, w_2 for constant $w_1 = \bar{w}_1$; this was not possible with model (1) or (1a) which did not permit independent variation of each factor. The slope of the curve $x = \varphi(\bar{w}_1, w_2)$ is the marginal productivity, which was not defined in relation to model (1) or (1a); φ_2' is discontinuous at the point of intersection with the expansion path, where φ is not differentiable [2].

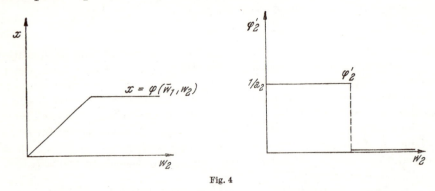

Fig. 4

When defined in this way, with the w_i as independent variables to be interpreted as available quantities of inputs, the limitational model presents itself as a limiting case of a production model with factor substitution. Formally, the factor combination for given x—and thus the expansion path—is no longer uniquely determined by the technology but may be thought of as the result of a cost minimization, only the least-cost combina-

---

[1] Strictly speaking, it might be argued that the concept of technical efficiency should apply only to models in which the input variables represent quantities consumed; the quantities in which inputs happen to be available have nothing to do with the technology. Suppose that, in a limitational model of the type (1 c), the first and the second input represent the services of two fixed factors, available in given amounts $w_1 = \bar{v}_1$, $w_2 = \bar{v}_2$ where $\bar{v}_1/\bar{v}_2 < a_1/a_2$ (cf. point B in Fig. 3). This means that, since it is impossible within the given technology to use both capacity factors to the full, waste cannot be entirely avoided so that no efficient point as defined here exists. This is a somewhat odd conclusion in view of the fact that there exists a perfectly respectable optimum, characterized in terms of model (1a) by $v_1 = a_1 x = \bar{v}_1$ and $v_2 = a_2 x < \bar{v}_2$, which is efficient in the sense referring to the latter model.

[2] The marginal productivity of w_2 is

$$\varphi_2'(\bar{w}_1, w_2) = \begin{cases} 1/a_2 & \text{for } 0 < w_2 < \bar{w}_1 a_2/a_1 \\ 0 & \text{for } \bar{w}_1 a_2/a_1 < w_2. \end{cases}$$

tion is wholly insensitive to factor prices changes[1]. In Fig. 3, point C will always be the least-cost point; this follows directly from the fact that it is the only point on $ABCDE$ which is technologically efficient in the sense of involving no waste of inputs, but it can also be derived by the familiar geometric procedure of finding the least-cost point as the point of "tangency" with an isocost line[2].

The two models (1a) and (1c) are of course equivalent and lead to the same results with respect to optimization. However, for reasons stated above, (1a) is better suited to production and cost analysis, and has the advantage of greater simplicity. The purpose of (1c) is to make the model appear as a (limiting) case of factor substitution, but to apply the analytical apparatus of marginal substitution analysis to a case which is primarily characterized by the total absence of substitution (and by non-differentiability) is hardly convenient and only tends to blur the fundamental difference between production with dependent and independent input variables.

B. Discontinuous Substitution: The Linear Production Model

1. The Production Relations

(a) Now suppose, as a generalization of the case of limitational inputs, that the same output can be produced in a finite number (N) of physically distinct sub-processes or "activities" using wholly or in part the same inputs, each of the activities being characterized by constant coefficients of production a_{ik}, so that we have

$$v_{ik} = a_{ik}\lambda_k \qquad (i = 1, 2, \ldots, m; \ k = 1, 2, \ldots, N)$$

where v_{ik} is the amount of input no. i required to produce λ_k units of output in the k'th activity. If the activities are not alternative in the sense of being mutually exclusive but can be applied simultaneously, total output becomes $x = \sum_k \lambda_k$ and the amount of input no. i required is $v_i = \sum_k v_{ik}$ so that the model of the joint process can be written

$$x = \sum_{k=1}^{N} \lambda_k$$

$$v_i = \sum_{k=1}^{N} a_{ik}\lambda_k \qquad (i = 1, 2, \ldots, m) \tag{2}$$

where the "activity levels" must be required to be non-negative:

---

[1] Cf. SAMUELSON (1947), p. 72.

[2] Total cost of the two variable inputs is $c = q_1 w_1 + q_2 w_2$, geometrically represented by a family of straight lines (isocosts) with c as a parameter. Through any point of the isoquant passes an isocost line and the one passing through C will represent minimum cost for the given x regardless of the prices, since the slope of the isocosts is negative for any set of positive factor prices.

$$\lambda_k \geq 0 \qquad (k = 1, 2, \ldots, N) \tag{3}$$

since negative output is economically meaningless. Finally, for those inputs which represent the services of fixed factors we have capacity limitations of the form

$$v_s \leq \bar{v}_s. \tag{4}$$

This is the *linear production model*. Its linear properties are implicit in the underlying assumption that the input-output proportions v_{ik}/λ_k are independent of the particular λ_k (constant coefficients of production within each activity) as well as of all other activity levels[1]; the activities are then divisible and additive so that total output and total inputs are linear functions of the activity levels.

(2) may be thought of as a parametric production model, from which the parameters λ_k can in some cases be eliminated so that we are left with a linear model in the output and input variables only. This is always possible when $N < m + 1$ and the activity vectors are linearly independent. For example, the model

$$x = \lambda_1 + \lambda_2$$
$$v_1 = 3\lambda_1 + 4\lambda_2$$
$$v_2 = \lambda_1 + 2\lambda_2$$
$$v_3 = 4\lambda_1 + 3\lambda_2$$

can be transformed by solving two of the equations—the second and the third, say—for the two activity levels:

$$\lambda_1 = v_1 - 2v_2$$
$$\lambda_2 = -\tfrac{1}{2}v_1 + \tfrac{3}{2}v_2$$

and substituting in the other two equations to get

$$x = \tfrac{1}{2}v_1 - \tfrac{1}{2}v_2$$
$$v_3 = \tfrac{5}{2}v_1 - \tfrac{7}{2}v_2$$

which is the production model in terms of x and the v_i only[2]. To these linear equations we must add the non-negativity requirements (3) which now become linear inequalities in v_1 and v_2:

$$(\lambda_1 =) \quad v_1 - 2v_2 \geq 0$$
$$(\lambda_2 =) -\tfrac{1}{2}v_1 + \tfrac{3}{2}v_2 \geq 0$$

or

$$\tfrac{1}{3}v_1 \leq v_2 \leq \tfrac{1}{2}v_1,$$

---

[1] The latter implies that the activities neither support nor impede one another. If the extent to which one activity is used has an adverse effect on the labour coefficient of another activity—for example, because labour productivity is lowered by noise, smoke, etc. resulting from the former activity—total labour input v_i cannot be written as a linear function of the two activity levels; we have instead

$$v_i = a_{i1}\lambda_1 + a_{i2}(\lambda_1) \cdot \lambda_2$$

where a_{i2} is a function of λ_1.

[2] In FRISCH's terminology, v_3 here appears as a "shadow factor" which depends on v_1 and v_2. Cf. Ch. V, A below.

i.e., the ratio of v_2 to v_1 is bounded by the relative technical coefficients in the two activities.

In the general case, however, no such elimination is possible and the model has to be considered in the general form (2).

Within each separate activity no factor substitution is possible since the inputs are limitational with respect to output λ_k. The model (2)–(4) as a whole, however, permits of substitution in that the same total output can be produced by an infinite number of linear combinations of the activities (except in the special case $N=1$, which is the limitational production model with constant coefficients); for given $x=\bar{x}$, the isoquant is defined in parametric form by

$$v_i = \sum_{k=1}^{N} a_{ik}\lambda_k \qquad (i = 1, 2, \ldots, m)$$

where the parameters—the activity levels—must be non-negative and satisfy

$$\sum_{k=1}^{N} \lambda_k = \bar{x}.$$

For example, in the two-factor, two-activity case

$$a_{11} = 2 \qquad a_{12} = 4$$
$$a_{21} = 6 \qquad a_{22} = 2$$

one unit of output may be produced either by employing the first activity only (point A in $Fig.$ $5a$, the activity being represented by the half-line from the origin through A), i.e., by the factor combination $v_1 = 2$, $v_2 = 6$, by the second activity only (point B, where $v_1 = 4$, $v_2 = 2$), or by joint application of the two activities with positive activity levels adding up to 1. Such a convex combination represents an intermediate point on the line segment AB, which is therefore the isoquant corresponding to $x=1$. For example, if unit output is distributed evenly between the two activities ($\lambda_1 = \lambda_2 = \frac{1}{2}$), the resulting factor combination—the total amounts of the two inputs consumed—will be that represented by point $P = (3, 4)$ in Fig. 5a, halfway between A and B[1]. The second activity requires more of v_1 and

---

[1] The equation of the straight line through A and B,

$$v_2 - a_{22} = \frac{a_{21} - a_{22}}{a_{11} - a_{12}} \cdot (v_1 - a_{12}),$$

is easily shown to be satisfied by the factor combination

$$v_1 = a_{11}\lambda_1 + a_{12}(1 - \lambda_1)$$
$$v_2 = a_{21}\lambda_1 + a_{22}(1 - \lambda_1).$$

Moreover, solving each of the latter equations for λ_1 in terms of v_1 and v_2 respectively, it can be shown that, if we require $\lambda_1 \geq 0$ and $\lambda_2 = 1 - \lambda_1 \geq 0$, this implies

$$a_{11} \leq v_1 \leq a_{12} \text{ and } a_{22} \leq v_2 \leq a_{21}$$

so that the range of feasible factor combinations is restricted to the segment AB, points beyond the extreme points A and B being incompatible with the non-negativity requirements.

An intermediate point may be thought of as the vector resulting from a linear combination of the activity vectors with the non-negative activity levels as coefficients:

less of v_2 per unit output than does the first; hence, increasing λ_2 at the expense of $\lambda_1\,(=1-\lambda_2)$ implies a substitution of v_1 for v_2 along the isoquant AB.

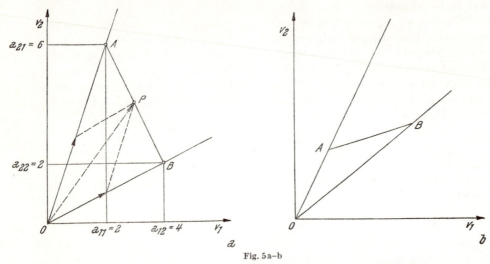

Fig. 5a–b

The half-line extending from the origin through any intermediate point on AB may be thought of as an activity possessing the same linear properties (fixed technical coefficients, constant returns to scale) as the two "elementary" activities from which it has been derived: proportionate variation in λ_1 and λ_2 implies that x, v_1, and v_2 vary in the same proportion, cf. (2). Hence all isoquants may be constructed as radial projections of AB.

When AB is downward sloping as in Fig. 5a, no point on the isoquant is technically inefficient: the same amount of output can be produced with less of one input only by using more of the other. In the case of AB being positively inclined (or parallel to one of the axes), however, only one of the activities is efficient; thus in *Fig. 5b*, whereas all points on the isoquant AB are feasible, a complete ordering of the points can be established[1] to show that point B is clearly inefficient, and so are all intermediate points. Only the activity containing A will represent efficient production.

In the case of more than two activities, the isoquants will be convex polygons—radial projections of one another—such as $ABCDE$ in *Fig. 6*. The boundary segments AB, BC, etc. represent linear combinations of two activities yielding a total output of one unit. An interior point of the polygon employs three or more activities; all such points are obviously inefficient, as well as points on the segments AE and ED. For example, E cannot be

$$\begin{bmatrix} x \\ v_1 \\ v_2 \end{bmatrix} = \lambda_1 \cdot \begin{bmatrix} 1 \\ a_{11} \\ a_{21} \end{bmatrix} + \lambda_2 \cdot \begin{bmatrix} 1 \\ a_{12} \\ a_{22} \end{bmatrix},$$

cf. the parallelogram of forces. Point P of Fig. 5a is illustrated in this way in the factor diagram, the coefficients being $\lambda_1 = \lambda_2 = \frac{1}{2}$.

[1] Cf. Ch. II above.

an efficient point since it is possible to produce a unit of output with less
of both inputs (e.g. with the factor combination Q, using the second and

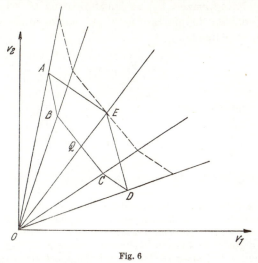

Fig. 6

the third activities), or to produce more than one unit with the factor combination E (cf. the dotted isoquant which represents an $x > 1$). Thus we are
left with points on the segments AB, BC, and CD so that the set of efficient
combinations for $x = 1$ is represented by the broken line segment $ABCD$,
which is convex to the origin and along which the marginal rate of substitution $\mid dv_2/dv_1 \mid_{x=\bar{x}}$ is piecewise constant, discontinuous at the kinks. Any
efficient point is a combination of at most two activities (two being the number of inputs), and the fifth activity—that containing E—does not enter into
any such combination.

This is readily extended to cases of three or more inputs. The isoquants,
as defined by (2) for $\sum_k \lambda_k = \bar{x}$, $\lambda_k \geq 0$, will be convex sets in factor space
and the set of efficient points for $x = \bar{x}$ will be those facets which "face the
origin" and which are negatively inclined to each of the axes. An efficient
point will represent a linear combination of at most m activities[1].

Thus, in a linear model of type (2)–(3), the factor combination for given
level of output can be varied within limits so that there is a range of economic choice—the "region of substitution"—bounded by the "extreme"
activities; in Fig. 6, the feasible combinations are represented by all points
on or between the first and the fourth activity half-lines (OA and OD).
However, such substitution as is possible is *discontinuous* in the sense that
the range of substitution is generated by a finite number of elementary
linear activities, any feasible overall factor combination being some "aver-

---

[1] Maximizing $x = \Sigma \lambda_k$ for given v_i, i.e., subject to $\Sigma a_{ik}\lambda_k = v_i$ $(i = 1, 2, \ldots, m)$ where
the parameters $v_1, v_2, \ldots, v_m$ represent any given feasible factor combination, is a linear
programming problem whose solution will have at most m positive λ_k, m being the number
of side conditions. This result follows directly from the fundamental theorem of linear programming, cf. Appendix 1.

age" of the combinations required for the same level of output by each of the activities that are employed[1][2].

The model is characterized by constant returns to scale. A given but arbitrary set of activity levels $\lambda_k = \lambda_k{}^0$ (≥ 0) corresponds to a point P in factor space with coordinates

$$v_i{}^0 = \sum_{k=1}^N a_{ik}\lambda_k{}^0$$

and belonging to the isoquant

$$x^0 = \sum_{k=1}^N \lambda_k{}^0.$$

Multiplying the activity levels by the same number μ leads to a new point

---

[1] It might be argued that "discontinuous substitution" is hardly an appropriate term for a case where the relative factor combination can in fact be varied continuously within limits. However, the classical case of *continuous* substitutability, as defined with reference to a continuous and differentiable production function $x = x(v_1, v_2, \ldots, v_m)$, is characterized by infinitesimal variation of the marginal rate of substitution being possible (again within limits), i.e., by smoothly curved isoquants, whereas in the linear case the (efficient) isoquants are piecewise linear—a property which reflects discontinuities in the marginal productivities—so that the least-cost combination will respond in a discontinuous manner to changes in the factor prices (cf. below).

[2] The linear model of production was developed after World War II in connexion with the technique of linear programming, due to DANTZIG and KOOPMANS—cf. KOOPMANS (1951), esp. Chs. II–III—and applied to the theory of the firm by DORFMAN (1951). However, the idea of factor substitution under a linear technology characterized by a finite number of linear activities was anticipated by ZEUTHEN (1942), p. 66 (English ed. (1955), pp. 64 ff.), who had previously pointed out that several "methods of production"—i.e., sets of technical coefficients—may be used simultaneously to produce the same commodity, cf. ZEUTHEN (1928), pp. 34 and 38, and (1933), pp. 18 ff.

In a somewhat different context—namely, in showing that concave pieces of an isoquant belonging to a homogeneous production function with continuous substitutability cannot be efficient—the idea of combining activities linearly was used, and illustrated by the parallelogram of forces, by GLOERFELT-TARP (1937), pp. 257 f., and generalized to the multi-product case by SCHMIDT (1939), pp. 293 ff. Discontinuities in the marginal rate of substitution were treated in a general way—not restricted to the linear case—by SAMUELSON (1947), Ch. 4 (cf. below).

For an empirical example of discontinuous substitution see CHARNES, COOPER and FARR (1953). The machine shop considered in this example manufactures several products (which are interdependent in that they share the services of the plant's fixed equipment), but each product can be produced in a finite number of linear activities because alternative kinds of fixed equipment can be used. For example, the first of the products listed can be manufactured on any one of four different types of screw machines. For the fifth product there are two types of screw machines which can do each other's jobs and two types of grinders, and each of the four combinations defines an activity.

It must be pointed out that the linear programming model characterized by a finite number of activities also applies to allocation problems which can scarcely be described as cases of discontinuous substitution as defined here. For example, the problem may be to find the least-cost combination of given materials (blending ingredients) which gives a product (blend) of specified composition; each input (ingredient) defines an activity whose coefficients represent the composition of the ingredient in terms of the constituents required in the blend. Models of this kind are dealt with in Ch. VIII, C below.

$$v_i = \sum_{k=1}^{N} a_{ik} \mu \lambda_k^0 = \mu \cdot v_i^0$$

on the isoquant

$$x = \sum_{k=1}^{N} \mu \lambda_k^0 = \mu \cdot x^0$$

so that the isoquants are radial projections of one another. Accordingly, the *elasticity of production*, i.e., the elasticity of x with respect to a proportionate variation in all inputs (or, which comes to the same thing, with respect to the scale factor μ), is unity everywhere; for $dv_i/v_i = v_i^0 d\mu / \mu v_i^0 = d\mu/\mu$ $(i=1, 2, \ldots, m)$ we have

$$\varepsilon = \frac{dx}{dv_i} \cdot \frac{v_i}{x} = \frac{dx}{d\mu} \cdot \frac{\mu}{x} = x^0 \cdot \frac{\mu}{\mu \cdot x^0} = 1$$

identically in μ. Setting $x^0 = 1$ so that $x = \mu$, the parameter μ can be thought of as representing the level of a derived activity whose technical coefficients are the v_i^0 and which is geometrically illustrated by the half-line from the origin through P.

There may, of course, be zeroes among the coefficients a_{ik}, i.e., the various activities produce the same output using wholly or in part different inputs. This is likely to be the rule rather than the exception: while it is perhaps difficult to see why it should be possible to produce the same output only by a finite number of sets of technical coefficients (though these activities can be used jointly), it is easy to imagine cases in which each activity is characterized by using an input which is specific to it (e.g. a fixed factor whose services require the variable inputs in technologically fixed proportions). In such cases each new activity adds a new dimension to the factor space.

(b) We have defined the v_i as quantities consumed of the various inputs. Alternatively, we may ask the question: How much is it possible to produce with given *available* quantities of the inputs, $w_1, w_2, \ldots, w_m$? The answer is given by

$$x = \text{Max} \sum_{k=1}^{N} \lambda_k \text{ subject to } (v_i =) \sum_{k=1}^{N} a_{ik} \lambda_k \leq w_i \text{ and } \lambda_k \geq 0$$

$$(i = 1, 2, \ldots, m; \ k = 1, 2, \ldots, N).$$

The maximization procedure will ensure uniqueness so that we can write the resulting x as a single-valued production function with the w_i as independent variables:

$$x = \varphi(w_1, w_2, \ldots, w_m). \tag{5}$$

(In the special case of $N = 1$, i.e., limitationality, the function will have the form

$$x = \text{Min}(w_i/a_i) \qquad (i = 1, 2, \ldots, m),$$

cf. above.)

The isoquant map of (5) differs somewhat from that of (2), the fundamental model from which (5) has been derived. In the two-factor case of Fig. 6, inefficient points of the polygon $ABCDE$ are now ruled out because of the maximization of x underlying (5). On the other hand, since w_1 and w_2 are independent variables, points outside the range of substitution must also be considered. If the inputs are available in amounts corresponding to point P in *Fig. 7* (the technology being the same as in Fig. 6), the maximum output that can be produced is the same as in point D; the amounts consumed of the two inputs are also the same, that part of w_1 which corresponds to DP being wasted. Hence the isoquant consists not only of the broken line $ABCD$ but also of half-lines from A and D parallel to the axes, as shown in Fig. 7. The efficient part of the isoquant is $ABCD$ as before; points on the half-lines (e.g. P), though representing maximal x for given w_1 and w_2, are inefficient in that it is possible to produce the same x with less of one input and the same amount of the other[1].

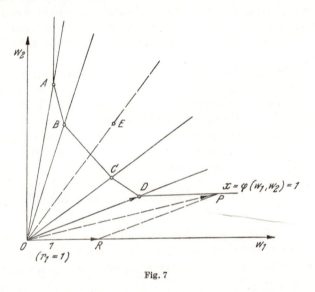

Fig. 7

The production function (5) can be illustrated in terms of production curves instead of isoquants. In the two-factor case of Fig. 7, x will vary with w_2 for constant $w_1 = \bar{w}_1$ in the fashion shown in *Fig. 8a*, the kinks

---

[1] In the terminology of linear programming, such a point may be constructed as a linear combination of one of the "structural" activities and a "slack activity" which produces nothing and "consumes" the input which is in abundant supply, the technical coefficients being 1 and 0. Point P of Fig. 7, for example, is the resultant of the vectors OD (the fourth structural activity) and OR, the slack activity of the first input. Measuring the level of the slack activity, r_1, in terms of that part of w_1 which is not used, we have

$$\begin{bmatrix} w_1 \\ w_2 \end{bmatrix} = \lambda_4 \cdot \begin{bmatrix} a_{14} \\ a_{24} \end{bmatrix} + r_1 \cdot \begin{bmatrix} 1 \\ 0 \end{bmatrix}$$

which for $\lambda_4 = 1$ and $r_1 = OR (= DP)$ gives the coordinates of P, cf. the parallelogram of forces shown in Fig. 7.

occurring whenever the line $w_1 = \overline{w}_1$ intersects an activity half-line[1]. (Only that part of the curve which lies between M and N, that is, in the region of substitution, would have been included if the inputs had been the amounts consumed, v_i.) The corresponding marginal productivity curve,

$$\varphi_2' = \delta\varphi(w_1,\, w_2)\,/\,\delta w_2$$

for $w_1 = \overline{w}_1$, is a step function with discontinuities at the same points, cf. *Fig. 8b*. At point N, output cannot be increased further—the isoquant becomes vertical—and the marginal productivity of w_2 drops to zero.

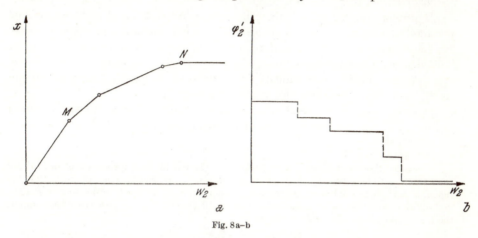

Fig. 8a–b

Model (5) is interesting only with a view to comparing with the smoothly curved isoquants and production curves of continuous substitution. The function φ is not differentiable at the strategic points and therefore unwieldy; optimization problems with discontinuous substitution are more easily solved when the technical side conditions have the linear form of (2), in which case the computational technique of linear programming is applicable.

2. Optimization Problems

(a) Returning to the model (2)–(3), let us consider the optimization problems with which the firm is confronted.

The least-cost factor combination for given level of output—assuming as usual that all inputs can be bought in unlimited amounts at constant prices—is determined by minimizing the cost expression $c = \Sigma q_i v_i$ subject to (2)–(3) for $x = \overline{x}$, that is, by minimizing the linear function

---

[1] In the region between, say, the first and the second activity half-lines we have for $w_1 = \overline{w}_1$

$$x = \lambda_1 + \lambda_2$$
$$\overline{w}_1 = a_{11}\lambda_1 + a_{12}\lambda_2$$
$$w_2 = a_{21}\lambda_1 + a_{22}\lambda_2;$$

eliminating λ_1 and λ_2, x becomes a linear function of $\overline{w}_1$—which is constant—and w_2. Between the second and the third activities we also get a linear function but the coefficients will be different; and so on.

$$c = \sum_{i=1}^{m} q_i \sum_{k=1}^{N} a_{ik} \lambda_k = \sum_{k=1}^{N} \left(\sum_{i=1}^{m} q_i a_{ik} \right) \lambda_k = \sum_{k=1}^{N} c_k \lambda_k \qquad (6)$$

(where c_k is cost per unit produced in the k'th activity) subject to the linear side condition

$$\sum_{k=1}^{N} \lambda_k = \bar{x} \qquad (7)$$

for $\lambda_k \geq 0$ $(k = 1, 2, \ldots, N)$.

A problem of this type cannot be solved by traditional methods of constrained maximization (elimination of dependent variables or, more generally, Lagrange's method of undetermined multipliers)[1] but will have to be solved as a problem of linear programming, though in this case a trivial one: the optimal solution is obviously to produce the required $\bar{x}$ units of output in that activity which has the lowest unit cost c_k.[2]

Thus the least-cost point will be one of the "corners" of the efficient boundary of the isoquant[3]. In the two-factor case shown in *Fig. 9* there are

----

[1] Cf. Appendix 1.

[2] Consider the minimization of (6) subject to (7) and (3) as a formal problem of linear programming, to be solved by the simplex method (see Appendix 1). Then we know from the fundamental theorem of linear programming that a single activity will suffice to give an optimal solution since there is only one linear side condition, (7). Now suppose that, for example, the first activity has the lowest unit cost:

$$c_1 < c_k \qquad (k = 2, 3, \ldots, N).$$

Choosing the first activity as a basis, we solve (7) for λ_1 and substitute in (6) to get total cost as a function of the non-basic variables $\lambda_2, \lambda_3, \ldots, \lambda_N$:

(i) $\qquad\qquad \lambda_1 = \bar{x} - \lambda_2 - \ldots - \lambda_N$

(ii) $\qquad\qquad c = c_1 \bar{x} + (c_2 - c_1)\lambda_2 + \ldots + (c_N - c_1)\lambda_N.$

The coefficients in (ii) are, by assumption, all positive; it follows that, since $\lambda_2, \lambda_3, \ldots, \lambda_N$ cannot be negative, c is a minimum for all these variables equal to zero so that the basic solution

$$\lambda_1 = \bar{x}, \; c = c_1 \bar{x}$$

represents minimum cost.

Any other feasible solution (non-negative set of values of activity levels λ_k) will represent higher cost so that the solution is also unique.

Had we assumed instead that $c_1 = c_2$, all other $c_k > c_1$, the coefficient of λ_2 would have been zero and the value of c would not have been affected by λ_2 becoming positive (but $\leq \bar{x}$ or λ_1 would become negative). In this case any non-negative combination of the first and the second activities for which $\lambda_1 + \lambda_2 = \bar{x}$ would represent minimum cost; but it would still be possible to attain minimum cost by using one activity only (the first or the second).

Thus it is a *sufficient* condition for minimum that the coefficients of the non-basic variables in (ii)—the "simplex coefficients"—are all non-negative. This condition, known as the *"simplex criterion"*, is also a *necessary* condition (except for the special case of "degeneracy," which cannot occur here) since the solution could be improved if one of the coefficients had been negative.

[3] Only efficient points need be considered since cost can never be a minimum when it is possible to produce the same amount of output with less of one input and not more of any other (cf. Ch. II above). On the other hand, the inclusion of inefficient points—which are often difficult to detect and sort out beforehand in more complicated problems—will do no harm since they will automatically be thrown out during the simplex calculations, just as they are ruled out by geometric considerations in the simple two-factor example.

three such points—A, B, and C—where only one activity is used and the marginal rate of substitution is discontinuous. A corner minimum of this

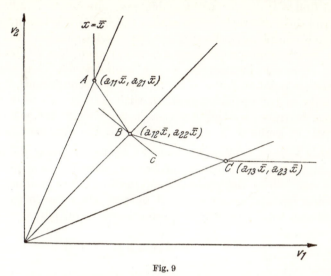

Fig. 9

type is characterized not by a marginal equality but by a pair of inequalities. The necessary and sufficient condition for, say, B to represent minimum cost is

$$\frac{a_{21} - a_{22}}{a_{11} - a_{12}} \leq -\frac{q_1}{q_2} \leq \frac{a_{22} - a_{23}}{a_{12} - a_{13}}, \tag{8}$$

i.e., the ratio of factor prices must be greater than or equal to the marginal rate of substitution along BC and less than or equal to that along AB; the slope of the family of isocost lines $c = q_1 v_1 + q_2 v_2$ must be between that of AB and that of BC in order for an isocost to be "tangent" to the isoquant at B as shown in Fig. 9. Condition (8) is seen to be equivalent to

$$c_2 \leq c_1, \quad c_2 \leq c_3\ [1]; \tag{9}$$

unit cost at point B (of the second activity) must be less than or equal to unit cost at the adjacent corners A and C. In the special case where $c_2 = c_1$,

---

[1] These inequalities, and thus also (8), are of course equivalent to the simplex criterion. Minimizing $c = c_1 \lambda_1 + c_2 \lambda_2 + c_3 \lambda_3$ subject to $\lambda_1 + \lambda_2 + \lambda_3 = \bar{x}$, using the second activity as a basis (in the manner shown above), we get

$$c = c_2 \bar{x} + (c_1 - c_2) \lambda_1 + (c_3 - c_2) \lambda_3$$

where the criterion for $\lambda_2 = \bar{x}$, $\lambda_1 = \lambda_3 = 0$ to be an optimal solution is that the coefficients of λ_1 and λ_3 are ≥ 0, cf. (9).

These coefficients,

$$\frac{\delta c}{\delta \lambda_1} = c_1 - c_2, \qquad \frac{\delta c}{\delta \lambda_3} = c_3 - c_2,$$

represent the marginal cost of moving from the least-cost solution in the direction of λ_1 and λ_3 respectively (transferring a unit of output from the second to the first or the third activity, keeping total output constant), i.e., the extra cost incurred by not using that factor combination which is cheapest at the given prices.

$c_2 < c_3$, so that the isocosts are parallel to AB, any point of the segment AB represents minimum cost so that the solution is not unique, but it is still possible to find a least-cost point which uses one activity only (A or B).

As long as the price ratio keeps within the limits (8), the optimal factor combination is completely insensitive to factor price changes: it is only when the isocost lines become steeper than AB or less steep than BC that the optimum point responds, jumping to A or C respectively. Thus the response of the least-cost point to factor price changes is discontinuous. Specifically, the sensitivity of the optimum position with respect to one of the prices, say, q_1, may be traced by treating q_1 as a parameter; for q_2 constant, (8) defines a pair of limits to q_1 within which B is the least-cost point[1].

(b) With $\bar{x}$ as a parameter, the locus of least-cost points defines the expansion path as a straight line in the non-negative factor space (except in the special case where the optimal combination is not unique), the same activity being optimal at all levels of output when the factor prices are constant. Hence total cost will be proportional to output,

$$c = \left(\sum_{i=1}^{m} q_i a_{it} \right) \cdot \lambda_t = c_t \cdot x,$$

activity no. t being the cheapest. Since total revenue is also proportional to x when the price of the product is constant,

$$r = p \cdot x,$$

no finite maximum of profit exists and an optimal level of output cannot be determined on the basis of revenue and cost when all inputs are available in unlimited amounts.

However, the latter assumption is not a realistic one: in short-run optimization problems there will always be one or more scarce inputs representing the services of fixed factors of production, i.e., there will be additional restrictions of the type (4) which impose upper bounds on certain inputs and thus also on the level of output. (The coefficients of such inputs in the cost function will usually be zero since total cost is not affected by the degree of capacity utilization as such.) Hence the factor combination determined by minimizing (6) subject to (7) and (3) represents a feasible least-cost solution only in so far as it respects the capacity restrictions (4); if it does not, the problem has to be reformulated with (4) as additional side conditions. Let the M first inputs ($M < m$) be the services of fixed factors with capacities $\bar{v}_1, \bar{v}_2, \ldots, \bar{v}_M$ (and prices $q_1 = q_2 = \ldots = q_M = 0$). Then the problem of cost minimization for given $x = \bar{x}$ is that of minimizing total variable cost of the current inputs $v_{M+1}, v_{M+2}, \ldots, v_m$:

$$c = \sum_{i=M+1}^{m} q_i v_i = \sum_{k=1}^{N} c_k \lambda_k \left(\text{where } c_k = \sum_{i=M+1}^{m} q_i a_{ik} \right) \tag{10}$$

subject to the non-negativity requirements (3) and to

---

[1] We might go further and study the succession of optimal solutions obtained as q_1 varies from zero upwards. This would be an example of "parametric programming" where some of

$$(x =) \sum_{k=1}^{N} \lambda_k = \bar{x} \qquad (7)$$

and

$$(v_i =) \sum_{k=1}^{N} a_{ik} \lambda_k \leq \bar{v}_i \,^1 \qquad (i = 1, 2, \ldots, M). \qquad (11)$$

For parametric $\bar{x}$, the solution to this linear programming problem determines the factor combination as depending on the level of output, i.e., the expansion path and the corresponding cost function. The optimal level of output can then be determined by maximizing total revenue minus total cost.

The optimal value of x could have been determined directly, without using the expansion path and the cost function as an intermediate stage, by maximizing total (gross) profit

$$z = px - c = \sum_{k=1}^{N} (p - c_k) \lambda_k = \sum_{k=1}^{N} z_k \lambda_k \qquad (12)$$

(where $z_k = p - c_k$ is unit gross profit in the k'th activity) subject to (3) and (11) [2]. This procedure is all the more expedient in cases of this type as the marginal cost function derived from the linear model turns out to be discontinuous.

(c) Linear programming problems cannot be solved analytically in the sense that it is possible to express the optimal values of the variables (here, the λ_k and thus also x and the v_i) in terms of the unspecified parameters a_{ik}, c_k or z_k (i.e., p and the q_i), and $\bar{v}_i$. The simplex method is essentially a method of numerical solution: the simplex criterion gives a necessary and sufficient criterion for optimality, but whether it is satisfied by a particular basic solution or not depends upon the numerical values of the parameters.

The optimization procedures may be illustrated by a numerical example with two factors. Let the production technology be represented by four activities with input coefficients as follows:

---

the strategic coefficients are treated as unspecified parameters (in the present case, the c_k, which are all affected by a change in q_1).

[1] By the fundamental theorem of linear programming, at most $M + 1$ positive variables—structural and/or slack variables—are required for an optimal solution, $M + 1$ being the number of linear side conditions (7)–(11). There are N structural variables λ_k ($k = 1, 2, \ldots, N$) and M slack variables r_i ($i = 1, 2, \ldots, M$), the latter defined by rewriting the capacity restrictions (11) in the form

$$\sum_{k=1}^{N} a_{ik} \lambda_k + r_i = \bar{v}_i \qquad (i = 1, 2, \ldots, M)$$

where r_i (≥ 0) is to be interpreted as idle capacity of the i'th fixed factor ($r_i = \bar{v}_i - v_i$, i.e., factor services available minus services used). The occurrence of a particular positive slack variable r_i in the optimal solution indicates that the i'th fixed factor is not an effective limit to output (at the given price constellation) so that the corresponding side condition might as well have been left out.

[2] The optimal solution, as determined by linear programming methods, will involve not more than M (in the absence of degeneracy, just M) positive variables.

a_{ik}	$k =$			
	1	2	3	4
$i =$ 1	4	3	2	1
2	2	3	4	6

The first input represents the services of a fixed factor, free ($q_1 = 0$) but available up to the capacity limit $\bar{v}_1 = 24$ only; the second is a variable factor whose price is $q_2 = 2$. Then the least-cost combination is determined by minimizing $c = q_2 v_2 = 2 v_2$ for $x = \bar{x}$ and $v_1 \leq \bar{v}_1$ (or $v_1 + r = \bar{v}_1$ where the slack variable r represents unused capacity), i.e., by minimizing

$$c = 4\lambda_1 + 6\lambda_2 + 8\lambda_3 + 12\lambda_4 \tag{13}$$

subject to

$$\lambda_1 + \lambda_2 + \lambda_3 + \lambda_4 = \bar{x} \tag{14}$$

$$4\lambda_1 + 3\lambda_2 + 2\lambda_3 + \lambda_4 + r = 24 (= \bar{v}_1) \tag{15}$$

where all variables—including the slack variable r, whose coefficient in the cost function is zero—are required to be ≥ 0.

This is a (parametric) linear programming problem, readily solved by the simplex method[1].

The first step is to select a basis of two variables, corresponding to the number of linear side conditions. The first activity is the cheapest and will obviously be preferred for $\bar{x} \leq \bar{v}_1/a_{11} = 6$, the maximum quantity that can be produced in activity no. 1 alone. This suggests the choice of (λ_1, r) as an initial basis. Solving (14)–(15) for these variables and substituting in (13) we have

$$\lambda_1 = \bar{x} - \lambda_2 - \lambda_3 - \lambda_4$$
$$r = (24 - 4\bar{x}) + \lambda_2 + 2\lambda_3 + 3\lambda_4 \tag{16}$$
$$c = 4\bar{x} + 2\lambda_2 + 4\lambda_3 + 8\lambda_4.$$

The third of these equations shows that cost will be a minimum for $\lambda_2 = \lambda_3 = \lambda_4 = 0$ because c is an increasing function of the three variables so that the simplex criterion is satisfied[2]. Hence the least-cost solution is

$$\lambda_1 = \bar{x}, \quad r = 24 - 4\bar{x}, \quad c = 4\bar{x} \tag{17}$$

as expected, only the solution is restricted to the interval $0 \leq \bar{x} \leq 6$, as r becomes negative (i.e., the solution is no longer feasible) for $\bar{x} > 6$. Only the first activity is used; the solution is the same as the one we should have found by minimizing (13) subject to (14) only. Marginal cost ($=$ variable unit cost) is $dc/d\bar{x} = 4$, the coefficient of $\bar{x}$ in the third equation of (16).

---

[1] The numerical procedure in the following corresponds to that outlined in Appendix 1 except for the slight complication due to the presence of the unspecified parameter $\bar{x}$, which makes the problem a case of parametric programming.

[2] The simplex criterion for a basic feasible solution to be a minimum requires that the "simplex coefficients"—the coefficients of the non-basic variables in the equation giving c as a function of these variables—shall be non-negative (cf. Appendix 1). Had one of the coefficients, say that of λ_2 in the third equation of (16), been zero, this would have indicated that the solution was optimal but not unique since a marginal increase in λ_2 (at the expense of λ_1) would not affect c.

For $\bar{x} > 6$ the capacity limit becomes effective and we must look for another solution in which r is $= 0$. A natural choice of a new basis is (λ_1, λ_2), i.e., some combination of the cheapest and the second-cheapest activity, λ_2 having now replaced r as a non-zero variable. Solving for λ_1, λ_2, and c in terms of the other three variables—this can be done either from (13)–(15) as before or, more easily, from (16)—we get

$$\lambda_1 = (24 - 3\bar{x}) + \lambda_3 + 2\lambda_4 - r$$
$$\lambda_2 = (4\bar{x} - 24) - 2\lambda_3 - 3\lambda_4 + r$$
$$c = (12\bar{x} - 48) + 0\lambda_3 + 2\lambda_4 + 2r$$

where again the simplex criterion is satisfied so that

$$\lambda_1 = 24 - 3\bar{x}, \quad \lambda_2 = 4\bar{x} - 24, \quad c = 12\bar{x} - 48 \tag{18}$$

—a combination of the two activities—is a least-cost solution within the interval where it is feasible, i.e., for $\lambda_1 = 24 - 3\bar{x} \geq 0$ and $\lambda_2 = 4\bar{x} - 24 \geq 0$, or $6 \leq \bar{x} \leq 8$. Within this interval marginal cost is again constant, now equal to $dc/d\bar{x} = 12$. At $\bar{x} = 6$, where the two intervals overlap, the two solutions (17) and (18) coincide and only one activity is used.

Proceeding in this fashion, the following least-cost solutions for higher levels of output emerge:

$$\lambda_2 = 24 - 2\bar{x}, \quad \lambda_3 = 3\bar{x} - 24, \quad c = 12\bar{x} - 48 \text{ for } 8 \leq \bar{x} \leq 12 \tag{19}$$
$$\lambda_3 = 24 - \bar{x}, \quad \lambda_4 = 2\bar{x} - 24, \quad c = 16\bar{x} - 96 \text{ for } 12 \leq \bar{x} \leq 24 \tag{20}$$

where marginal cost is 12 and 16 respectively. For $\bar{x} > 24$ no non-negative solution exists; $\bar{x} = 24$ is the maximum output that can be produced, as was evident from the outset since $\bar{v}_1/a_{14} = 24$, the fourth activity having the lowest coefficient of v_1 per unit of output.

These results are illustrated in Fig. 10a–c. *Fig. 10a* shows that the expansion path in the factor diagram is $0ABCD$, where $0A$ corresponds to solution (17), AB to (18), etc. The same result could also have been arrived at by geometrical considerations, recalling that for $q_1 = 0$ the isocosts are parallel to the v_1 axis. The cost function (total variable cost) is piecewise linear, cf. *Fig. 10b* (where A', B', etc. correspond to A, B, etc. in Fig. 10a) and marginal cost is a step function as shown in *Fig. 10c*.

The optimal point on the expansion path $0ABCD$ can then be determined by comparing marginal revenue (= the product price, p) with the marginal cost function as shown in Fig. 10c. For $p = 15$, for example, profit is a maximum at C where $x = \lambda_3 = 12$, marginal cost being

$$\frac{dc}{dx} = \begin{cases} 16\,(>15) \text{ for } x > 12 \\ 12\,(<15) \text{ for } x < 12 \,. \end{cases}$$

This result could have been arrived at directly by maximizing the profit function

$$z = 15x - 2v_2 = 11\lambda_1 + 9\lambda_2 + 7\lambda_3 + 3\lambda_4 \tag{21}$$

subject to the capacity restriction (15) for $\lambda_k \geq 0$ and $r \geq 0$. Solving (15) for λ_3 and substituting in (21) we have

$$\lambda_3 = 12 - 2\lambda_1 - \tfrac{3}{2}\lambda_2 - \tfrac{1}{2}\lambda_4 - \tfrac{1}{2}r$$
$$z = 84 - 3\lambda_1 - \tfrac{3}{2}\lambda_2 - \tfrac{1}{2}\lambda_4 - \tfrac{7}{2}r,$$

(22)

where z is a decreasing function of the non-basic variables so that (22) represents a maximum solution for $\lambda_3 = 12$, all other variables equal to zero[1].

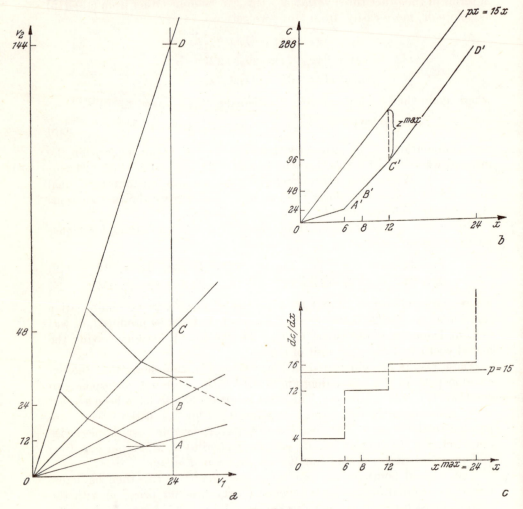

Fig. 10a–c

More generally, for parametric p the optimal level of output is that which makes the profit function

$$z = px - 2v_2 = (p-4)\lambda_1 + (p-6)\lambda_2 + (p-8)\lambda_3 + (p-12)\lambda_4$$

a maximum subject to (15). Obviously for profit to be positive at all we must have $p > 4$. With λ_1 as a basis the solution is

$$\lambda_1 = 6 - \tfrac{3}{4}\lambda_2 - \tfrac{1}{2}\lambda_3 - \tfrac{1}{4}\lambda_4 - \tfrac{1}{4}r$$
$$z = (p-4)\cdot 6 + (\tfrac{1}{4}p - 3)\lambda_2 + (\tfrac{1}{2}p - 6)\lambda_3 + (\tfrac{3}{4}p - 11)\lambda_4 + (-\tfrac{1}{4}p + 1)r$$

(23)

[1] For a *maximum*, the simplex criterion requires all simplex coefficients to be *non-positive*.

where all coefficients in the profit expression are ≤ 0 for $4 \leq p \leq 12$ so that the solution

$$x = \lambda_1 = 6, \quad z = (p - 4) \cdot 6$$

(cf. point A in Fig. 10a) represents maximum profit for $4 < p \leq 12$. In a similar manner we find for higher values of p that the solutions

$$x = \lambda_3 = 12, \quad z = (p - 8) \cdot 12 \quad \text{for } 12 \leq p \leq 16$$
$$x = \lambda_4 = 24, \quad z = (p - 12) \cdot 24 \quad \text{for } 16 \leq p$$

(cf. points C and D in Fig. 10a) are optimal solutions. The supply function thus developed corresponds to the marginal cost function illustrated in Fig. 10c. For $p = 12$, where the price is exactly equal to marginal cost, the solution is not unique, as reflected in the zero coefficients of λ_2 and λ_3 in the profit expression in (23); any point on ABC (Fig. 10a) represents maximum profit. Similarly, for $p = 16$ either C or D or any intermediate point will be optimal.

(d) In determining the optimal level of output by maximizing profit subject to capacity limitations we have implicitly determined a shadow price for each capacity factor, representing the marginal value to the firm of the factor's services.

In (22), for example, which is an optimal solution for $\lambda_1 = \lambda_2 = \lambda_4 = r = 0$, the coefficient of r in the profit expression—the simplex coefficient of the slack variable—is $\delta z / \delta r = -\frac{7}{2}$. This means that profit will be reduced by $\frac{7}{2}$ if the utilization of the services of the fixed factor (which was fully utilized at the optimal point, $r = 0$) is reduced by one unit. Conversely, if one more unit of the scarce input—e.g. an additional machine hour—had been available, its being used would raise profit by the same amount. Thus the value of the marginal unit of the fixed factor's services, in the sense of forgone profit or marginal contribution to profit, is $\left| \dfrac{\delta z}{\delta r} \right| = \dfrac{7}{2}$.

The interpretation is clear if we return to the original equations (21) and (15), which are of course equivalent to (22). Starting from the optimal solution, where $\lambda_1 = \lambda_2 = \lambda_4 = r = 0$, the effect of a marginal increment in r is given by the differentials

$$dz = 7 \cdot d\lambda_3$$
$$2 \cdot d\lambda_3 + dr = 0$$

from which

$$\frac{\delta z}{\delta r} = \frac{\delta z}{\delta \lambda_3} \cdot \frac{\delta \lambda_3}{\delta r} = 7 \cdot \left(-\frac{1}{2} \right) = -\frac{7}{2},$$

i.e., the marginal effect on output in the third activity times unit profit. If capacity $\bar{v}_1$ rather than r is treated as an independent parameter, we have instead (with $r = 0$)

$$dz = 7 \cdot d\lambda_3$$
$$2 \cdot d\lambda_3 = d\bar{v}_1,$$

i.e., the effect of a unit increment in capacity is

$$\frac{\delta z}{\delta \bar{v}_1} = \frac{\delta z}{\delta \lambda_3} \cdot \frac{\delta \lambda_3}{\delta \bar{v}_1} = \frac{7}{2}.$$

In the case of two fixed factors and two structural activities where the problem is to maximize

$$z = z_1 \lambda_1 + z_2 \lambda_2 \tag{24}$$

subject to

$$\begin{aligned} a_{11}\lambda_1 + a_{12}\lambda_2 + r_1 \quad &= \bar{v}_1 \\ a_{21}\lambda_1 + a_{22}\lambda_2 \quad + r_2 &= \bar{v}_2 \end{aligned} \tag{25}$$

(where z_1 and z_2 are unit gross profits and r_1 and r_2 are slack variables), suppose that profit is a maximum when the first activity is used up to the limit set by the first capacity so that the solution for (λ_1, r_2),

$$\begin{aligned} \lambda_1 &= \bar{v}_1/a_{11} - (a_{12}/a_{11})\lambda_2 - (1/a_{11})r_1 \\ r_2 &= \bar{v}_2 - \bar{v}_1 a_{21}/a_{11} + (a_{21}a_{12}/a_{11} - a_{22})\lambda_2 + (a_{21}/a_{11})r_1 \quad (26) \\ z &= z_1\bar{v}_1/a_{11} - (z_1 a_{12}/a_{11} - z_2)\lambda_2 - (z_1/a_{11})r_1, \end{aligned}$$

is optimal for $\lambda_2 = r_1 = 0$, a situation illustrated by points P and P' in *Fig. 11a–b*.[1] This will be so if the simplex coefficients are non-positive; in the present case they are strictly negative, the solution being unique.

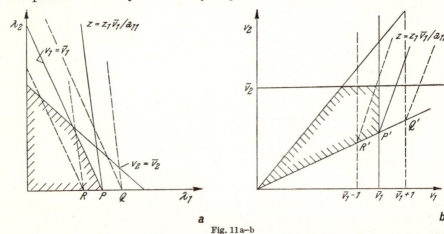

a   b

Fig. 11a–b

As to the first fixed factor which is fully utilized in the optimum position ($r_1 = 0$), a marginal increase in the capacity parameter $\bar{v}_1$ (λ_2 and r_1 still $= 0$) will affect λ_1 and z positively and diminish r_2, cf. the constant terms in (26). The new solution is thus feasible when the increase in $\bar{v}_1$ is small enough

---

[1] *Fig. 11a* illustrates the problem (24)–(25) as it stands in a (λ_1, λ_2) diagram; *Fig. 11b* shows the activities and the capacity limits in the factor plane. The shaded areas represent the set of feasible solutions. Since Fig. 11a assumes that

$$-z_1/z_2 < -a_{11}/a_{12} < -a_{21}/a_{22}$$

(the slopes of the iso-profit and capacity lines), it follows that the first activity—requiring relatively more of v_1—is the half-line from the origin through P' in Fig. 11b and that the family of profit lines in Fig. 11b (as determined by solving (25) for λ_1 and λ_2 in terms of $\bar{v}_1 - r_1 = v_1$ and $\bar{v}_2 - r_2 = v_2$ and substituting in (24)) are positively inclined in the manner shown in the figure.

not to make r_2 negative[1]; it is also optimal since the simplex coefficients are not affected. Thus, by the third equation of (26), the marginal value of the services of the first fixed factor is

$$\frac{\delta z}{\delta \bar{v}_1} = \frac{z_1}{a_{11}} \qquad \left(= \frac{\delta z}{\delta \lambda_1} \cdot \frac{\delta \lambda_1}{\delta \bar{v}_1}, \text{ cf. (24)–(25)}\right),$$

which is numerically equal to the simplex coefficient of r_1,

$$\frac{\delta z}{\delta r_1} = -\frac{z_1}{a_{11}} \qquad \left(= \frac{\delta z}{\delta \lambda_1} \cdot \frac{\delta \lambda_1}{\delta r_1}, \text{ cf. (24)–(25)}\right),$$

the interpretation being that a marginal increase in r_1 (unutilized capacity) for given $\bar{v}_1$ is equivalent to a reduction of capacity for constant $r_1 = 0$. The effects of a positive and a negative shift in $\bar{v}_1$ on the optimal position are represented in Fig. 11 by a move from P to Q or R (or from P' to Q' or R').

Fig. 11 also suggests that a (small) marginal change in the second capacity $\bar{v}_2$ will not change the equilibrium position. At the given prices v_2 is not a bottleneck factor ($r_2 > 0$), so an increase in $\bar{v}_2$ will add nothing to profit but will merely result in a corresponding increase in unutilized capacity r_2; the factor is not effectively scarce but is a free good. This is confirmed by (26): the new solution where r_2 and $\bar{v}_2$ have been increased by the same amount leaves everything else as it was, so it is feasible and optimal for $\lambda_2 = r_1 = 0$ and we have

$$\frac{\delta z}{\delta \bar{v}_2} = 0.$$

Correspondingly, the simplex coefficient of r_2 may be said to be zero since r_2, being a basic variable, does not enter into the third equation of (26).

Thus the shadow prices associated with the fixed factors are both non-negative, being numerically equal to the simplex coefficients of the corresponding slack variables in the optimal basic solution[2]; the shadow price is zero if the factor is not fully utilized ($r_i > 0$), positive if the factor is effectively scarce (or zero in the special case where the simplex coefficient of r_i happens to be zero so that the solution is not unique). This result can easily be shown to hold for any number of fixed factors (M) and structural activities (N)[3] where total gross profit

$$z = \sum_{k=1}^{N} z_k \lambda_k \tag{12}$$

---

[1] In the special ("degenerate") case of r_2 being $= 0$ in the initial optimum position—i.e., when the two capacity lines intersect at P (or P')—$\bar{v}_1$ cannot be increased without r_2 becoming negative, unless r_1 is increased similarly. This means that a marginal increase in the capacity $\bar{v}_1$ will add nothing to profit; the same applies to an increase in $\bar{v}_2$.

[2] In numerically specified models, where $\bar{v}_i$ does not occur explicitly as a parameter in the solution but is concealed in the constant terms in (26), we cannot calculate $\delta z / \delta \bar{v}_i$ directly from the solution, but we can always find the simplex coefficient $\delta z / \delta r_i$.

[3] For a proof see Appendix 1.

is maximized subject to non-negativity requirements $\lambda_k \geq 0$ $(k = 1, 2, \ldots,$ $N)$ and to the capacity limitations

$$\sum_{k=1}^{N} a_{ik}\lambda_k \leq \bar{v}_i \text{ or } \sum_{k=1}^{N} a_{ik}\lambda_k + r_i = \bar{v}_i \qquad (i = 1, 2, \ldots, M) \qquad (11)$$

where the slack variables r_i must also be ≥ 0 or the inequalities would be reversed.

By the duality theorem of linear programming, the shadow prices determined by solving this problem correspond to the optimal solution of the "dual" problem of minimizing the linear function

$$g = \sum_{i=1}^{M} \bar{v}_i y_i \qquad (27)$$

subject to

$$\sum_{i=1}^{M} a_{ik}y_i \geq z_k \text{ or } \sum_{i=1}^{M} a_{ik}y_i - s_k = z_k \qquad (k = 1, 2, \ldots, N) \qquad (28)$$

where the y_i as well as the slack variables s_k must be non-negative[1]. The optimal value of y_i is numerically equal to the simplex coefficient of r_i in the optimal solution to the maximization problem, and s_k is numerically equal to the simplex coefficient of λ_k; conversely, λ_k and r_i are equal to the simplex coefficients of s_k and y_i respectively. The maximum value of z is equal to the minimum of g.

Consequently the y_i, as determined by solving the dual problem, can be identified with the shadow prices associated with the respective factor services v_i. The slack variables s_k also have an economic interpretation: by (28) we have

$$-s_k = z_k - \sum_{i=1}^{M} a_{ik}y_i$$

which is "net" profit per unit of output in the k'th activity, defined as gross profit ($z_k = p - c_k$ where c_k is variable cost per unit) minus the opportunity cost of utilizing the fixed factors, using the shadow prices as accounting prices per unit of factor service used. Unit net profit in this sense is non-positive since $s_k \geq 0$ in the optimal solution.

This result provides an alternative method of optimal planning: in solving the dual problem we have implicitly solved the problem of maximizing gross profit subject to (11), i.e., determined the optimal activity levels λ_k. The interpretation of the dual is that only those activities should be used whose unit net profit is zero: $s_k = 0$ means that s_k, being a non-basic variable in the optimum solution, has a positive simplex coefficient which is equal to λ_k [2]. All other activities have negative net profit and this indicates that it

---

[1] The duality theorem, which was first applied to shadow pricing in the firm by DORFMAN (1951), pp. 45 ff., is stated in Appendix 1. For a simple proof of the theorem see, e.g., DANØ (1960), pp. 109–113.

[2] Assuming uniqueness, the simplex coefficients of the non-basic variables will be strictly negative in the maximum problem, strictly positive in the dual (minimum) problem. If zeroes occur in one of the problems, the other will be "degenerate."

does not pay to use them; a positive s_k in the optimal basis—i.e., $-s_k < 0$—will have the simplex coefficient 0 so that $\lambda_k = 0$ [1]. In the optimum position, total gross profit z will just cover the total imputed cost of the fixed factors, g.

It follows that short-run production planning on a "full-cost" basis—with total cost equal to the given selling price as a criterion for profitability—will lead to maximum profit if and only if the cost of using the fixed factors is based on factor prices coinciding with the shadow prices y_t.

C. Discontinuities in General

A more general, but less operational treatment of discontinuities in the production function has been given by SAMUELSON[2]. The analysis is based upon a production function of the same general type as (1 c) or (5) above—i.e., the inputs are defined as *independent* variables to be interpreted as available quantities, and $x = \varphi(w_1, w_2)$ is the maximum output that can be produced with any given combination of w's—only no assumption of linearity (homogeneity) or of a finite number of activities is made. All that is assumed is that an isoquant contains only a finite number of points which do not possess continuous partial derivatives; the isoquant is continuous but kinked. This will be so at a point where the marginal productivities and thus also their ratio, the marginal rate of substitution, are discontinuous.

In *Fig. 12*, let $P = (w_1, w_2)$ be such a point. Q and R are neighbouring points on the same isoquant $x = \bar{x}$ so that we have

$$\varphi(w_1, w_2) = \varphi(w_1 - \Delta w_1, w_2 + \Delta w_2) = \varphi(w_1 + \Delta w_1, w_2 - \Delta w_2) = \bar{x}$$

for Δw_1 and $\Delta w_2 > 0$. At P, the marginal productivity of w_1 is discontinuous, being $\varphi_1'^L$ and $\varphi_1'^S$ for respectively negative and positive small changes in w_1 where $\varphi_1'^L > \varphi_1'^S$ as shown in the figure (L stands for "largest", S for "smallest"); similarly, for small partial variations in w_2 we have $\varphi_2'^L > \varphi_2'^S$.

Now for a move from P to Q we have

$$\Delta x = - \varphi_1'^L \Delta w_1 + \varphi_2'^S \Delta w_2 = 0. \tag{29}$$

P represents a cheaper (or at least not more expensive) factor combination than Q if we also have for the resulting change in total variable cost

$$\Delta c = - q_1 \Delta w_1 + q_2 \Delta w_2 \geq 0. \tag{30}$$

Similarly, moving from P to R and assuming that this does not reduce cost, we have

---

[1] In the two-activity case above, where $\lambda_1 > 0$ and $\lambda_2 = 0$ in the optimal solution, unit net profits are

$$(-s_1 =) z_1 - y_1 a_{11} - y_2 a_{21} = z_1 - \frac{z_1}{a_{11}} \cdot a_{11} - 0 \cdot a_{21} = 0$$

and

$$(-s_2 =) z_2 - y_1 a_{12} - y_2 a_{22} = z_2 - \frac{z_1}{a_{11}} \cdot a_{12} - 0 \cdot a_{22} < 0,$$

cf. the negative simplex coefficient of λ_2 in (26).

[2] Cf. SAMUELSON (1947), pp. 70–76.

$$\Delta x = \varphi_1'^S \Delta w_1 - \varphi_2'^L \Delta w_2 = 0 \tag{31}$$

$$\Delta c = q_1 \Delta w_1 - q_2 \Delta w_2 \geq 0 . \tag{32}$$

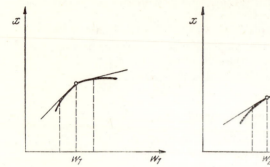
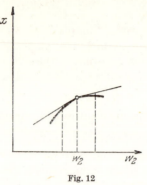
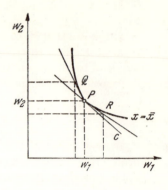

Fig. 12

Eliminating Δw_2 from (29)–(30) and from (31)–(32) respectively we find that the necessary and sufficient condition for P to be a local least-cost point (not necessarily unique) is

$$\frac{\varphi_1'^S}{\varphi_2'^L} \leq \frac{q_1}{q_2} \leq \frac{\varphi_1'^L}{\varphi'_2{}^S} . \tag{33}$$

Thus the factor combination P is insensitive to factor price changes as long as the price ratio keeps within the bounds set by the left-hand and right-hand marginal rates of substitution at P [1]. At the upper and lower critical values of, say, q_1 for constant q_2 the response of w_1 and w_2 to marginal changes in q_1 will be discontinuous.

If the isoquant is assumed to be convex to the origin so that the isocost line $c = q_1 w_1 + q_2 w_2$ which is "tangent" to $x = \bar{x}$ at P does not intersect the isoquant, P represents not only a local minimum of c but also a minimum "in the large"—a *global* optimum—for values of q_1 and q_2 satisfying (33).

Condition (33) is seen to include continuity as a special case: for

$$\varphi_1'^S = \varphi_1'^L = \varphi_1' \text{ and } \varphi_2'^S = \varphi_2'^L = \varphi_2'$$

(33) becomes the familiar marginal equality[2]

$$\frac{q_1}{q_2} = \frac{\varphi_1'}{\varphi_2'} .$$

The extreme case of discontinuity is that of limitational inputs. Let the production function be

$$x = \varphi(w_1, w_2) = \mathrm{Min}\,(w_1/a_1,\ w_2/a_2),$$

cf. (1 c). At a point where $x = w_1/a_1 = w_2/a_2$ the left- and right-hand partial derivatives are

---

[1] In the case of m inputs, (33) generalizes to $(m-1)$ pairs of inequalities

$$\frac{\varphi_1'^S}{\varphi_i'^L} \leq \frac{q_1}{q_i} \leq \frac{\varphi_1'^L}{\varphi_i'^S} \qquad (i = 2, 3, \ldots, m),$$

each pair derived in a similar manner for compensating variations in w_1 and w_2, w_1 and w_3, etc.

[2] Cf. Ch. IV, D.

$$\varphi_1'^L = 1/a_1, \quad \varphi_1'^S = 0,$$
$$\varphi_2'^L = 1/a_2, \quad \varphi_2'^S = 0;$$

then (33) says that such a point is optimal for any factor price ratio from zero to infinity, i.e., for any set of non-negative prices.

SAMUELSON mentions no other particular examples of discontinuities. However, (33) also covers the more general case of a linear model with discontinuous substitution. In the three-activity case of Fig. 9 above, B is a least-cost point for factor prices satisfying (8) (which in turn is equivalent to the simplex criterion), as is obvious from geometrical considerations. Precisely the same condition for minimum cost emerges if the marginal productivities are calculated and applied to (33)[1].

---

[1] For small variations in the activity levels we have

$$\Delta x = \Delta \lambda_1 + \Delta \lambda_2 + \Delta \lambda_3$$
$$\Delta w_1 = a_{11} \Delta \lambda_1 + a_{12} \Delta \lambda_2 + a_{13} \Delta \lambda_3$$
$$\Delta w_2 = a_{21} \Delta \lambda_1 + a_{22} \Delta \lambda_2 + a_{23} \Delta \lambda_3.$$

Now, the left-hand derivative of the production function $x = \varphi(w_1, w_2)$ with respect to w_1 at point B is found by setting $\Delta w_2 = 0$ and $\Delta \lambda_3 = 0$ in the equations (the third activity is not used for efficient production to the left of B), eliminating $\Delta \lambda_1$ and $\Delta \lambda_2$:

$$\frac{\Delta x}{\Delta w_1} = \frac{a_{22} - a_{21}}{a_{22} a_{11} - a_{12} a_{21}} = \varphi_1'^L.$$

For $\Delta w_1 = 0$ and $\Delta \lambda_3 = 0$ we find for a positive partial variation in w_2:

$$\frac{\Delta x}{\Delta w_2} = \frac{a_{12} - a_{11}}{a_{12} a_{21} - a_{22} a_{11}} = \varphi_2'^S.$$

The ratio of the marginal productivities—i.e., the marginal rate of substitution along BA—is

$$\frac{\varphi_1'^L}{\varphi_2'^S} = \frac{a_{22} - a_{21}}{a_{11} - a_{12}}.$$

Similarly, along BC we have

$$\frac{\varphi_1'^S}{\varphi_2'^L} = \frac{a_{23} - a_{22}}{a_{12} - a_{13}}.$$

Substituting in (33), we get the conditions (8).

Chapter IV

Production Functions with Continuous Factor Substitution

A. Factor Substitution and the Isoquant Map

1. The Region of Substitution

(a) The "classical" type of factor substitution, long predominant in economic theory by virtue of its plausibility and analytical convenience, finds expression in a production function

$$x = x(v_1, v_2, \ldots, v_m), \tag{1}$$

assumed to possess continuous first- and second-order partial derivatives in the region of definition $v_i \geq 0$, $x \geq 0$. The underlying process must be technologically defined in such a way as to make the production function single-valued; when the existing technical knowledge leads to a multiple-valued function because a given set of input amounts v_i may be organized in different ways, this means that the process is incompletely specified. To obtain uniqueness in such cases it is usually assumed[1] that the factors are always organized in a technically optimal fashion so that the production function is defined as giving the maximum amount of output for any factor combination.

When the v_i are defined as inputs consumed in the process, points at which one or more marginal productivities $x_i' = \delta x / \delta v_i$ is negative are not excluded from the production function (1), only such a point is obviously inefficient in that a greater amount of output can be produced with less of the factor or factors in question, other inputs constant[2]. Hence, with a view to optimal allocation in production, we are particularly interested in points at which

$$x_i' \geq 0 \qquad (i = 1, 2, \ldots, m) \tag{2}$$

so that x is an increasing or at least non-decreasing function of all v_i. (1)–(2)

---

[1] Cf., e.g., CARLSON (1939), pp. 14 f.

[2] A negative marginal productivity implies that the input in question is actually harmful; leaving some of it idle would lead to greater output, thus apparently contradicting the definition of (1) as giving maximum output for any combination of inputs. The contradiction is resolved, however, if it is recalled that the v_i are, by definition, the amounts actually used, whether harmful or not. The alternative definition of the production function as giving maximum output for any combination of input amounts available, $x = \varphi(w_1, w_2, \ldots, w_m)$, is dealt with below.

define the *region of substitution* as the set of efficient points belonging to the production function. For marginal variations at a point belonging to (1) we have

$$dx = \sum_{i=1}^{m} x_i' \, dv_i;$$

if the point satisfies (2), output can be increased ($dx > 0$) only by using more of at least one input ($dv_i > 0$ for some i), and a decrease in some input ($dv_i < 0$ for some i) will have to be compensated by using more of another if output is to be kept at the same level, $dx = 0$, except in the limiting case where $x_i' = 0$ so that output is not affected by an infinitesimal decrease in v_i. In order for a range of factor substitutability to exist, the interior of the region of substitution—i.e., the region where the marginal productivities are strictly positive, $x_i' > 0$ for all i—must contain more than one point in factor space for given x so that mutually compensating variations in the v_i are possible, using more of some inputs and less of others.

(b) For given $x = \bar{x}$, (1) defines an isoquant surface along which the factor combination can be varied continuously. Infinitesimal variations along the isoquant will have to satisfy

$$(dx=) \sum_{i=1}^{m} x_i' \, dv_i = 0.$$

For partial variations in two of the factors, the others (as well as output) being constant, we have

$$\frac{\delta v_k}{\delta v_j} = -\frac{x_j'}{x_k'}$$

which is negative—i.e., dv_j and dv_k are of opposite signs—at a point in the interior of the region of substitution; on the boundary of the region, $\delta v_k / \delta v_j$ is zero (for $x_j' = 0$) or minus infinity (in the limit, for $x_k' \to 0$). The numerical value of the partial derivative at an efficient point,

$$\left| \frac{\delta v_k}{\delta v_j} \right| = \frac{x_j'}{x_k'} \, (\geq 0),$$

is the marginal rate of substitution, the ratio of small compensating variations. Taking v_m as the dependent variable in the equation of the isoquant

$$x(v_1, v_2, \ldots, v_m) = \bar{x},$$

the partial derivatives $\delta v_m / \delta v_i$ ($i = 1, 2, \ldots, m-1$) at a point of the isoquant represent the directions of the tangent plane at the point; they are continuous and differentiable functions of $v_1, v_2, \ldots, v_{m-1}$ in the interior of the region of substitution so that factor substitution—movements along the efficient part of the isoquant—is continuous, characterized by continuous variation in the marginal rates of substitution.

In the two-factor case this means that the isoquant

$$x(v_1, v_2) = \bar{x}$$

is smoothly curved and negatively inclined in the efficient region, the slope of the tangent being negative:

$$\frac{dv_2}{dv_1} = -\frac{x_1'}{x_2'} < 0 \qquad (3)$$

for x_1' and x_2' positive; cf. *Fig. 13*.

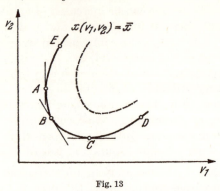

Fig. 13

In Fig. 13, D is an inefficient point: a horizontal move to the left will lead to a higher isoquant (i.e., $x_1' < 0$, whereas x_2' is positive since the tangent at D is positively inclined), or output $\bar{x}$ can be attained with less of both inputs; similarly, at point E we have $x_1' > 0$ but $x_2' < 0$. Only points on ABC are efficient, A and C being the points at which the marginal rate of substitution is ∞ ($x_1' > 0$, $x_2' = 0$) and 0 ($x_1' = 0$, $x_2' > 0$) respectively. At B both marginal productivities are positive.

For parametric $\bar{x}$, the production function is represented by a family of isoquants, as illustrated in the two-factor case by an isoquant map like that of *Fig. 14*. The region of substitution is bounded by $x_1' = 0$ and $x_2' = 0$, that is, by the loci $0M$ and $0N$ of horizontal and vertical tangents to the isoquants.

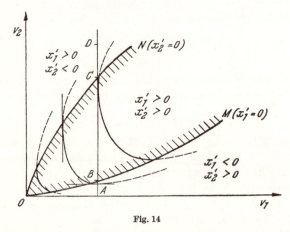

Fig. 14

The boundaries of the region of substitution may be straight lines, or they may be curvilinear and converge to a point representing maximum

output. It is also conceivable that both marginal productivities are positive
for any positive factor combination, in which case the region of substitution
consists of the entire positive region as bounded by the axes[1]—if the isoquants intersect or touch the axes, this means that it is possible to produce
the commodity using only one input—or that $x_1' = 0$ and $x_2' = 0$ do not
pass through the origin[2]. Examples of the latter kinds are illustrated by
Fig. 15 below.

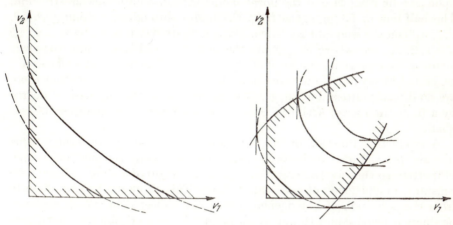

Fig. 15

In the case shown in Fig. 14, where the isoquants have no points in
common with the axes, the two factors are technically *complementary* in the
mild sense that it takes some of each input to get a positive output at all;
for efficient production the input combination is even more restricted,
namely to the region bounded by $0M$ and $0N$. (Perfect complementarity,
then, is the limiting case in which the region of substitution shrinks to a
curve in factor space so that $0M$ and $0N$ coincide and the inputs become
limitational.) In the interior of the region of substitution it is possible to
increase output marginally with more of one input and less of another, but
the boundaries of the region are characterized by marginal complementarity:
at a point on $0M$, where $x_1' = 0$, a marginal increase in v_1 will result in a
larger output only if accompanied by an increase in v_2.[3]

---

[1] In the special case of a linear-homogeneous Cobb-Douglas production function

$$x = k \cdot v_1{}^a \cdot v_2{}^{1-a}$$

(where k and α are constants, $0 < \alpha < 1$), x_1' and x_2' are positive for v_1 and v_2 positive and never
become zero except in the limit: for given $x = \bar{x}$ we have $x_1' \to 0$ (and $v_2 \to 0$) for $v_1 \to \infty$.

[2] Cf. HEADY and PESEK (1960), p. 901. The boundaries may even be negatively inclined
(viz. if $x_{12}'' = \delta^2 x / \delta v_1 \delta v_2 < 0$), cf. *op. cit.*, pp. 903 f.

[3] The latter concept of complementarity corresponds to that used by HEADY and PESEK
(1960), p. 902.

The term has been used by other authors in several other senses. A commonly used
definition is analogous to that used by EDGEWORTH and PARETO in the theory of consumers'
demand (cf. HICKS (1946), p. 42): two inputs are technically complementary if the second-
order cross derivative of the production function is positive, $x_{ij}'' > 0$, i.e., if an increase in
one input will raise the marginal productivity of the other (in the terminology of HEADY
and PESEK: the case of "positive interaction" between the inputs). Complementarity in this

2. Returns to Scale and Elasticities of Production

When the production function is homogeneous of degree one, that is, when

$$x(\mu v_1, \mu v_2, \ldots, \mu v_m) = \mu \cdot x(v_1, v_2, \ldots, v_m)$$

for any point on the production surface and for any (positive) value of μ, the isoquants are radial projections of one another. This case may be looked upon as a limiting case of the linear model with discontinuous substitution. The half-line in factor space from the origin through any point $(v_1^0, v_2^0, \ldots, v_m^0)$ on the isoquant $x = 1$ is a linear activity determined by $v_i = \mu \cdot v_i^0$ $(i = 1, 2, \ldots, m)$ where the v_i^0 are the technical coefficients and the scale factor $\mu = x$ represents the level of the activity[1]. The isoquant surfaces are generated by an infinite number of such elementary activities so that there are no discontinuities, but they can be approximated to any desired degree by a finite number of activities—the more activities, the smoother the isoquants will be.

A production function of this type is characterized by constant returns to scale: proportional variation in all inputs will change output in the same proportion so that average returns x/μ (or the ratio of x to each input) are constant, independent of the scale factor[2]. This property follows directly from the definition of homogeneity; it is also reflected in the elasticity of production being equal to one at all points. In general, for any given production function (1), let $(v_1^0, v_2^0, \ldots, v_m^0)$ be an arbitrary point on the isoquant $x = 1$ so that variations in the scale of output with the same relative

---

sense is not opposed to substitutability as defined above, cf. HICKS, *loc. cit.*; when $x = x(v_1, v_2)$ is homogeneous of degree one, x_{12}'' is positive everywhere in the region of substitution if the isoquants are convex, cf. Appendix 2.

Still another concept of complementarity is defined as the cross derivative of the firm's factor demand function being negative: $\delta v_i/\delta q_j < 0$. On rather restrictive assumptions only can it be proved that the inputs are complementary in demand if they are technically complementary in the sense that $x_{ij}'' > 0$ at the point of equilibrium, cf. FRISCH (1956), pp. 187 ff.

[1] The boundary of the region of substitution is generated by half-lines representing those activities in which $x_i' = 0$ for some i. For $\mu = 1/v_1$ (say), the definition of homogeneity leads to

$$x = x(v_1, v_2, \ldots, v_m) = v_1 \cdot x(1, v_2/v_1, \ldots, v_m/v_1)$$
$$= v_1 \cdot f(v_2/v_1, \ldots, v_m/v_1)$$

which is an equivalent way of writing the production function. The partial derivatives of the function in this form turn out to depend on the relative factor proportions v_i/v_1 only. It follows that x_i' is constant (e.g. = 0) for proportional factor variation. In Fig. 14, for example, OM and ON will be straight lines if the production function is homogeneous of the first degree.

[2] It must be emphasized again that the production function treated here refers to a process taking place within a *given plant*. The inputs associated with the fixed capital equipment represent the services of the respective items of equipment (e.g. machine hours per period), not these items themselves, which are assumed to be available in given numbers. The occurrence of constant returns to scale within the plant depends on the possibility of varying these inputs along with the other variable inputs (labour, etc.), i.e., of varying the utilization of the plant for constant relative factor proportions. Whenever the doubling of all inputs represents a mere repetition (in time or space), output will also be doubled. This problem, which must not be confused with the question of large-scale economies where plants of different sizes are compared, is treated in some detail below (Ch. VII).

factor proportions can be expressed by $v_i = \mu \cdot v_i{}^0$ ($i = 1, 2, \ldots, m$). Then the *elasticity of production*[1] at the point $(v_1, v_2, \ldots, v_m)$ may be defined as

$$\varepsilon = \frac{dx}{d\mu} \cdot \frac{\mu}{x},$$

that is, the elasticity of the production function

$$x = x(\mu \cdot v_1{}^0, \mu \cdot v_2{}^0, \ldots, \mu \cdot v_m{}^0)$$

with respect to the parameter μ. In general, the value of ε will depend on the v_i, i.e., on the scale factor μ and the relative factor proportions as given by the $v_i{}^0$; for small variations from a particular point, average returns to scale (x/μ) will be increasing, constant, or decreasing according as $\varepsilon \gtreqless 1$ at the point so that $d(x/\mu)/d\mu = (\mu \cdot dx/d\mu - x)/\mu^2 \gtreqless 0$. If the production function is homogeneous of the first degree we have $x(v_1, v_2, \ldots, v_m) = \mu \cdot x(v_1{}^0, v_2{}^0, \ldots, v_m{}^0)$ or $x = \mu$ identically and therefore $\varepsilon = 1$ at any point.

Alternatively, proportional factor variation may be expressed by $dv_i/v_i = k$ (constant) for all i and the elasticity of production may be defined accordingly as the elasticity of the production function with respect to any input, say, v_1:

$$\varepsilon = \frac{dx}{dv_1} \cdot \frac{v_1}{x} \qquad \text{(for } dv_i/v_i = k, \; i = 1, 2, \ldots, m).$$

The two definitions of ε are of course equivalent as $dv_i/v_i = v_i{}^0 d\mu/\mu \cdot v_i{}^0 = d\mu/\mu$.

The latter definition leads to an important theorem. The differential of x being $dx = \sum_i x_i{}' dv_i = k \cdot \sum_i v_i x_i{}'$ for $dv_i/v_i = k$, we have

$$\varepsilon = \frac{k \cdot \sum_i v_i x_i{}'}{k \cdot v_1} \cdot \frac{v_1}{x}$$

or

$$\varepsilon \cdot x = \sum_{i=1}^{m} v_i x_i{}' \qquad (4)$$

at any point on the production surface[2].

(4) holds for any production function (1); if the function is homogeneous of degree one, Euler's theorem says that $x = \sum v_i x_i{}'$ identically so that

---

[1] Also known as the "passus coefficient" (FRISCH), the "function coefficient" (CARLSON), or—in German—"der Ergiebigkeitsgrad" or "die Niveauelastizität" (SCHNEIDER); cf. SCHNEIDER (1934), pp. 9 ff., and (1958), p. 178.

[2] Theorem (4) also holds for proportional variation of a subset of inputs $v_k, \ldots, v_m$, the others being constant ($v_i = \bar{v}_1$ for $i = 1, 2, \ldots, k-1$) if ε is redefined accordingly as the elasticity of the function

$$\Phi(v_k, \ldots, v_m) = x(\bar{v}_1, \ldots, \bar{v}_{k-1}, v_k, \ldots, v_m)$$

with respect to v_k for $dv_i/v_i = $ constant ($i = k, \ldots, m$). The proof is the same. With this interpretation, (4) has been named the *Wicksell-Johnson theorem*, cf. SCHNEIDER (1934), pp. 19 ff.

$\varepsilon = 1$ [1]. More generally, for a production function which is homogeneous of order r so that multiplication of any factor combination by μ will multiply output by μ^r, the elasticity of production will be identically equal to r [2].

Another theorem follows directly from (4):

$$\varepsilon = \sum_{i=1}^{m} x_i' \cdot \frac{v_i}{x} = \sum_{i=1}^{m} \varepsilon_i \tag{5}$$

where ε_i is the elasticity of the production function with respect to v_i, the *partial elasticity of production*. In the case of first-degree homogeneity the partial elasticities add up to one.

3. The Convexity Assumption

(a) In order for any point in the region of substitution to represent minimum cost for some non-negative set of factor prices (cf. below), the isoquants must be assumed to be *convex* to the origin in the region. In the two-factor case this is simple enough: at any point (v_1, v_2) where $x_1' > 0$ and $x_2' > 0$, the curvature of the isoquant $x(v_1, v_2) = \bar{x}$ through the point must be positive:

$$\frac{d^2 v_2}{d v_1^2} = \frac{(x_2')^2 x_{11}'' - 2 x_1' x_2' x_{12}'' + (x_1')^2 x_{22}''}{-(x_2')^3} > 0, \tag{6}$$

that is, the slope $dv_2/dv_1 (<0)$ must be an increasing function of v_1. (6) is the condition for the isoquant to be convex *from below*, as seen from the axis of the independent variable v_1. Since the isoquant is negatively inclined in the region considered, it will also be convex from the other axis,

$$d^2 v_1 / d v_2^2 = (x_2'/x_1')^3 \cdot d^2 v_2 / d v_1^2 > 0,$$

and thus convex *to the origin* when (6) is satisfied. The interpretation of (6) is that the amount of v_2 saved by applying another unit of v_1 for x constant becomes smaller as the substitution proceeds: the marginal rate of substitution $|dv_2/dv_1| = -dv_2/dv_1 (>0)$ is a decreasing function of v_1 [3]. In

---

[1] For proof of Euler's theorem see, for example, Osgood (1925), pp. 121 f. (Conversely, since we know already that $\varepsilon = 1$ identically, (4) may be taken as proof of the theorem.)

[2] For a point $(v_1^0, v_2^0, \ldots, v_m^0)$ on $x = 1$ we have by the definition of r'th-order homogeneity

$$x = x(\mu v_1^0, \mu v_2^0, \ldots, \mu v_m^0) = \mu^r \cdot x(v_1^0, v_2^0, \ldots, v_m^0) \text{ or } x(\mu) = \mu^r$$

so that

$$\varepsilon = \frac{dx}{d\mu} \cdot \frac{\mu}{x} = r \cdot \mu^{r-1} \cdot \frac{\mu}{\mu^r} = r \text{ identically.}$$

The whole isoquant map can be constructed from a single isoquant (for example, $x = 1$) also for $r \neq 1$—for given μ, $x(\mu) = \mu^r = $ constant independent of the v_i^0, cf. above—only x does not vary in the same proportion as μ. Returns to scale are increasing or decreasing everywhere according as $r > 1$ or $r < 1$.

[3] Convexity is sometimes described as "*increasing* marginal rate of substitution", referring to $|dv_2/dv_1|$ as a function of v_2 (or $|dv_1/dv_2|$ as a function of v_1), cf. Allen (1938), p. 286.

Fig. 13, $|dv_2/dv_1|$ falls monotonically from infinity (at point A) to zero (point C) as v_1 is substituted for v_2 along the efficient part of the isoquant, ABC.

Whether or not condition (6) is actually satisfied in concrete cases is another matter, to be settled by empirical investigation. The possibility of alternating sections of convexity and concavity cannot be ruled out a priori[1]. Allowance must also be made for the limiting case of perfect substitutability where the isoquants are straight lines and thus not strictly convex, the marginal rate of substitution being constant so that $d^2v_2/dv_1{}^2 = 0$.[2]

More generally, a function is (strictly) convex from below in a region if linear interpolation between two arbitrary points always overestimates the

---

[1] A point in the region of substitution at which the isoquant is concave (P in the figure below) can never represent minimum cost at any given set of factor prices, though it is technically efficient in the sense that x_1' and x_2' are both positive and no other *particular* point on the isoquant is cheaper for *any* set of factor prices. (However, P may be a cost minimum if the isocost is concave to the origin, as it will be when the factor prices are increasing functions of the v_i.)

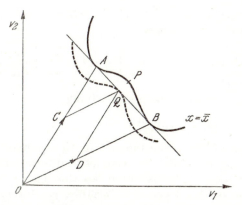

GLOERFELT-TARP (1937), pp. 257 f., assuming that the production function is homogeneous of degree one, has argued that such a point does not belong to the production function at all. Let the line AB be tangent to the isoquant $x=\bar{x}$ at A and B. Then, regarding $0A$ and $0B$ as linear activities each producing $\bar{x}$ units, any point on AB will represent a convex combination of $0A$ and $0B$ producing the same amount of output; and it is always possible to find such a point—for example, Q—which uses less of both inputs than does P. Hence P is inefficient and the concave part of the isoquant will be replaced by the line segment AB if—as GLOERFELT-TARP assumes—the production function is defined as giving *maximum* output for given v_1 and v_2. The points C and D, which combine to yield the point Q, must be interpreted as two physically distinct factor combinations used simultaneously; the corresponding two activities must be *additive* in order for the argument to hold.

[2] The isoquants will be straight lines if x is a linear function of v_1 and v_2 or, more generally, if the production function can be written in the form

$$f_1(x) \cdot v_1 + f_2(x) \cdot v_2 = F(x).$$

It is also conceivable that an otherwise curvilinear isoquant may have a flat section along which $|dv_2/dv_1|$ is constant, only the assumption of continuous second-order partial derivatives will then have to be abandoned. Let the isoquant in the figure above be linear between A and B. Then $d^2v_2/dv_1{}^2$ will be discontinuous at A, being positive to the left of A and zero to the right; and similarly at B.

actual value of the function[1]. With, say, v_m as the dependent variable, the equation of the isoquant

$$x(v_1, v_2, \ldots, v_m) = \bar{x}$$

can be written

$$v_m = f(v_1, v_2, \ldots, v_{m-1}).$$

The function f is convex from below in the region of substitution if the following inequality holds for any two distinct points $(v_1^0, \ldots, v_m^0)$ and $(v_1, \ldots, v_m)$:

$$(1 - \vartheta) \cdot f(v_1^0, \ldots, v_{m-1}^0) + \vartheta \cdot f(v_1, \ldots, v_{m-1})$$

$$> f\{[(1-\vartheta)v_1^0 + \vartheta v_1], \ldots, [(1-\vartheta)v_{m-1}^0 + \vartheta v_{m-1}]\}$$

for any positive fraction ϑ, $0 < \vartheta < 1$.[2] Rearranging we have

$$f(v_1, \ldots, v_{m-1}) - f(v_1^0, \ldots, v_{m-1}^0)$$

$$> \frac{f\{[v_1^0 + \vartheta(v_1 - v_1^0)], \ldots, [v_{m-1}^0 + \vartheta(v_{m-1} - v_{m-1}^0)]\} - f(v_1^0, \ldots, v_{m-1}^0)}{\vartheta}$$

where $f(v_1, \ldots, v_{m-1}) = v_m$ and $f(v_1^0, \ldots, v_{m-1}^0) = v_m^0$. In the limit (for $\vartheta \to 0$)[3] convexity is seen to imply

$$v_m > v_m^0 + \sum_{i=1}^{m-1} [f_i']^0 \cdot (v_i - v_i^0) \quad \left(= v_m^0 + \sum_{i=1}^{m-1} \left[\frac{-x_i'}{x_m'} \right]^0 \cdot (v_i - v_i^0) \right) \quad (7)$$

for any two distinct efficient points satisfying

$$x(v_1^0, \ldots, v_m^0) = x(v_1, \ldots, v_m) = \bar{x}.$$

Since the equation of the plane tangent to the isoquant at $(v_1^0, \ldots, v_m^0)$ is the corresponding equality, the geometric interpretation of (7) is that any other efficient point on the isoquant lies above the tangent plane—that is, in the direction of the dependent variable, v_m—except in the limiting case where $(x_m')^0 = 0$ so that $(f_i')^0 = -\infty$. As long as only efficient points are considered, (7) can be rearranged to give a similar inequality with any other input as dependent variable. In other words, if the isoquant function is convex in the region of substitution for some set of $m-1$ independent input variables—cf. (7)—the isoquant is convex in *all* factor directions, and therefore convex *to the origin*, in the efficient region. Rearranging (7) into the more symmetrical form

---

[1] This is the definition used by KUHN and TUCKER (1951)—cf. Appendix 1 below—except that the limiting case of linearity, where linear interpolation neither underestimates nor overestimates the function, is excluded. If it is to be allowed for, the inequality signs $>$ must be replaced by $\geq$ in the following.

[2] In the two-factor case the geometric interpretation is obvious: Between any two points, e.g. B and C in Fig. 13, the isoquant lies below the chord connecting the points (the line segment BC).

[3] For $\vartheta = 0$, the right-hand side of the inequality is an indeterminate form, $0/0$; for $\vartheta \to 0$ it approaches a limit which is the ratio of the derivatives of numerator and denominator with respect to ϑ. (Cf. OSGOOD (1925), pp. 208 f., and KUHN and TUCKER (1951), pp. 485 f.)

$$\sum_{i=1}^{m} (x_i')^0 \cdot (v_i - v_i^0) > 0 \qquad (x_i' \geq 0, \; i = 1, 2, \ldots, m), \tag{8}$$

which in equality form would represent the equation of the tangent plane at $(v_1^0, \ldots, v_m^0)$, convexity to the origin is seen to imply in geometric terms that all efficient points distinct from the point of tangency are "outside" the tangent plane. More precisely, the tangent plane—negatively inclined in all directions—at an arbitrary efficient point $(v_1^0, \ldots, v_m^0)$ on the iso-quant will be intersected by the line segment in factor space connecting the origin and any other efficient point $(v_1, \ldots, v_m)$ on the isoquant. For $m = 2$, (7) or (8) is equivalent to convexity in the sense of (6).[1]

Inequality (8), with the same geometric interpretation, also holds when $(v_1, \ldots, v_m)$ is an arbitrary *inefficient* point on the isoquant (still assuming that the arbitrary fixed point $(v_1^0, \ldots, v_m^0)$ is efficient). In order for a point $(v_1, \ldots, v_m)$ to represent the global minimum of $z = \Sigma (x_i')^0 v_i$ subject to $\bar{x} = x(v_1, v_2, \ldots, v_m)$, in which case it would be "inside" the tangent plane, it would have to satisfy the necessary conditions

$$(x_i')^0 - u \cdot x_i' = 0 \qquad (i = 1, 2, \ldots, m)$$

where u is a Lagrange multiplier. Since $(x_i')^0 \geq 0$ for all i, we have either $x_i' \geq 0$ or $x_i' \leq 0$ for all i at the point. Suppose the latter to be the case at a point $Q = (v_1^Q, \ldots, v_m^Q)$ belonging to the isoquant $x = \bar{x}$ and satisfying the necessary conditions for minimum z. Then select some other point $R = (v_1^R, \ldots, v_m^R)$ in factor space such that $v_i^R < v_i^Q$ for $i = 1, 2, \ldots, m$. For variations along the straight line through Q and R, the production function can be written in terms of a parameter t,

$$x = x(v_1, v_2, \ldots, v_m) = f(t),$$

where

$$v_i = v_i^R + t \cdot (v_i^Q - v_i^R) \quad (\gtreqless v_i^Q \text{ for } t \gtreqless 1), \; i = 1, 2, \ldots, m.$$

For $t = 1$ we have point Q where $v_i = v_i^Q$, $x = f(1) = \bar{x}$, whereas $t = 0$ corresponds to point R. The derivative of the function is

$$f'(t) = \sum_{i=1}^{m} \frac{\delta x}{\delta v_i} \cdot \frac{dv_i}{dt} = \sum_{i=1}^{m} (v_i^Q - v_i^R) \cdot x_i' (v_1, \ldots, v_m)$$

which for $t = 1$ becomes

---

[1] Expanding the isoquant function $v_2 = f(v_1)$ by Taylor's series (cf. ALLEN (1938), pp. 451 f.) we have

$$f(v_1) = f(v_1^0) + f'(v_1^0) \cdot (v_1 - v_1^0) + \tfrac{1}{2} \cdot f'' \cdot (v_1 - v_1^0)^2$$

where f'' $(= d^2v_2/dv_1^2)$ is taken at the point $v_1^0 + \vartheta(v_1 - v_1^0)$ for some positive fraction ϑ. It follows that, when (7) is satisfied, f'' will be positive at this intermediate point and therefore also at v_1^0 since the interval $v_1 - v_1^0$ can be made arbitrarily small and f'' is a continuous function. Hence, v_1^0 being an arbitrary point, (6) is satisfied everywhere in the interior of the region of substitution. Conversely, if (6) is satisfied everywhere, so are (7) and (8).

$$f'(1) = \sum_{i=1}^{m} (v_i^Q - v_i^R) \cdot (x_i')^Q < 0.$$

Assuming that $f(t) = 0$ for some $t = T < 1$—if this were not the case, a positive output would result from negative amounts of all inputs—it is obvious from a graph of the continuous function $x = f(t)$ that there exists an intermediate point in the region $T < t < 1$ where $f(t) = \bar{x}$, that is, another point $S = (v_1^S, \ldots, v_m^S)$ on the isoquant $x = \bar{x}$ such that $v_i^S < v_i^Q$ for all i. Hence we have

$$\sum_{i=1}^{m} (x_i')^0 v_i^S < \sum_{i=1}^{m} (x_i')^0 v_i^Q \, ;$$

in other words, a point Q characterized by $(x_i')^Q \leq 0$ $(i = 1, 2, \ldots, m)$ cannot represent the global minimum of z on the isoquant. The "best" point, therefore, is the best efficient point, $(v_1^0, \ldots, v_m^0)$. The tangent plane at this point has no other points—efficient or otherwise—in common with the isoquant, as expressed by (8).[1][2]

(b) The isoquant $x(v_1, \ldots, v_m) = \bar{x}$ cannot be illustrated geometrically by an isoquant map when $m > 2$, unless the substitution which is considered is restricted in such a way that only one input variable is capable of independent variation. The question is to what extent the characteristic features of the two-dimensional isoquant persist for variations of this kind.

For example, in the three-factor case we may wish to study compensating variations in v_1 and v_2 for constant $v_3 = \bar{v}_3$. The equation of the isoquant in terms of v_1 and v_2 is

$$x(v_1, v_2, \bar{v}_3) = \Phi(v_1, v_2) = \bar{x}$$

where $\Phi_1' = x_1'$ and $\Phi_2' = x_2'$. In the interior of the region of substitution belonging to the function Φ—the region where Φ_1' and Φ_2' are both positive—the isoquant will be downward sloping,

$$dv_2 / dv_1 = -\Phi_1' / \Phi_2' < 0.$$

Assuming that the three-dimensional isoquant is convex to the origin in the region where $x_i' > 0$ $(i = 1, 2, 3)$, (8) for $v_3 = v_3^0 = \bar{v}_3$ leads to

$$v_2 > v_2^0 + \left[-\frac{\Phi_1'}{\Phi_2'} \right]^0 \cdot (v_1 - v_1^0)$$

for any two points on $\Phi(v_1, v_2) = \bar{x}$ satisfying $\Phi_1' > 0$, $\Phi_2' > 0$, and $x_3' > 0$.

---

[1] If Q is a feasible point $(v_i^Q \geq 0$ for all $i)$, the origin may be used for R so that we have $v_i = t \cdot v_i^Q$; $f(t)$ will be zero for $t = 0$ (or for some t in the region $0 < t < 1$), and $f(t) = \bar{x}$ at some intermediate point. However, in order to prove *analytically* that (8) holds for *any* point $(v_1, \ldots, v_m)$ on the isoquant and therefore also for any *feasible* point, the possibility that there may exist infeasible points Q where $x_i' \leq 0$ for all i must be taken into account, though such points are devoid of economic meaning. In such cases the point R has to be selected differently in order to ensure $v_i^Q > v_i^R$ for all i.

[2] Since $(v_1^0, \ldots, v_m^0)$ may be any efficient point, (8) implies in the two-factor case that the isoquant is confined to the region bounded by the vertical and horizontal tangents and the efficient points (in Fig. 13, the tangents at A and C and the curve ABC).

It follows that the two-dimensional isoquant is convex in this region, but that convexity cannot be established everywhere in the region $\Phi_1' > 0$, $\Phi_2' > 0$ since (8) is not assumed to hold for $x_3' < 0$.[1]

Another kind of restricted variation is the substitution of v_1 against a homogeneous complex of v_2 and v_3, i.e., $v_3 = k \cdot v_2$ where k is a positive constant. Then we have

$$x(v_1, v_2, v_3) = x(v_1, v_2, kv_2) = \Phi(v_1, v_2) = \bar{x}$$

where $\Phi_1' = x_1'$ and $\Phi_2' = x_2' + kx_3'$. Again, the slope of the two-dimensional isoquant will be negative for $\Phi_1' > 0$, $\Phi_2' > 0$, but convexity cannot be established throughout this region since it is possible for Φ_2' to be positive for x_2' or $x_3' < 0$, in which case (8) is not assumed to be satisfied.

More general cases of substitution with fewer than $m - 1$ independent input variables along an isoquant can be examined in a similar manner. In general, the assumption (8) is not sufficient for the isoquant $\Phi = \bar{x}$ to be convex everywhere in the corresponding region of substitution.

4. The Elasticity of Substitution

The "degree of substitutability" between the inputs clearly depends on the curvature of the isoquant surface. In the two-factor case, the curvature of an isoquant can be expressed by the value of the second derivative d^2v_2/dv_1^2 (x constant), which—assuming convexity—is known to be positive at an efficient point. The flatter the isoquant is, the more we approach the limiting case of perfect substitutability where the isoquant is a straight line (and so no longer strictly convex) and d^2v_2/dv_1^2 becomes zero. The opposite extreme is the case where v_1 and v_2 must be used in a fixed proportion so that the range of substitution for given output—i.e., the efficient part of the isoquant—shrinks to a single point; as this limit is approached, the curvature tends to infinity because the interval of v_1 within which dv_2/dv_1 changes from $-\infty$ to 0 tends to zero.

[1] The general law of diminishing returns (cf. below) suggests, however, that x_3' will tend to be positive for sufficiently large v_1 and v_2; as the scale of output increases for v_3 constant, v_3 becomes relatively scarcer and x_3' is likely to increase so that the isoquant tends to be convex in the region.

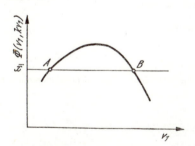 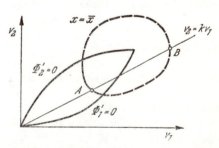

Diminishing returns with respect to proportional variation in v_1 and v_2 for constant v_3 also suggests that the isoquants $\Phi(v_1, v_2) = \bar{x}$ are closed curves (as stated by SCHNEIDER (1934), p. 25): expanding along a line $v_2 = kv_1$ (k constant), output will decline from a certain point so that the same isoquant is intersected twice, cf. points A and B in the figures above.

The second-order derivative d^2v_2/dv_1^2, however, has the disadvantage of being dependent on the units in which the inputs are measured. A more convenient measure of the degree of substitutability is provided by the *elasticity of substitution*, defined as the elasticity of the relative factor proportion along the isoquant with respect to a change in the marginal rate of substitution[1]:

$$s_{12} = \frac{d\,(v_2/v_1)}{d\,(-dv_2/dv_1)} \cdot \frac{(-dv_2/dv_1)}{v_2/v_1} = \frac{d\,(v_2/v_1)}{d\,(dv_2/dv_1)} \cdot \frac{dv_2/dv_1}{v_2/v_1}$$
$$= \frac{(v_2 - v_1 \cdot dv_2/dv_1) \cdot (-dv_2/dv_1)}{v_1 v_2 \cdot d^2v_2/dv_1^2} \tag{9}$$

for x constant. s_{12}, which is a pure number—positive in the interior of the region of substitution provided that the isoquant is convex, cf. (3) and (6)— is a function of v_1 and v_2, subject to $x(v_1, v_2) = \bar{x}$ and thus to $x_1'\,dv_1 + x_2'\,dv_2 = 0$.[2] The elasticity of substitution it not an appropriate tool for describing the shape of the whole isoquant—much less for characterizing a particular production function in its entirety—unless the shape of the isoquant happens to be such that s_{12} is constant along the curve[3]; the elasticity

---

[1] Cf. ALLEN (1938), pp. 341 ff.

[2] Evaluating s_{12} further, expressing dv_2/dv_1 and d^2v_2/dv_1^2 in terms of the derivatives of the production function (cf. (3) and (6) above), the elasticity becomes

$$s_{12} = \frac{-x_1' x_2' (v_1 x_1' + v_2 x_2')}{v_1 v_2 [(x_2')^2 x_{11}'' - 2x_1' x_2' x_{12}'' + (x_1')^2 x_{22}'']},$$

cf. ALLEN, *op. cit.*, p. 342.

If the production function is homogeneous of degree one, this reduces to

$$s_{12} = \frac{x_1' \cdot x_2'}{x \cdot x_{12}''},$$

using Euler's theorem; cf. ALLEN, *op. cit.*, p. 343. In this case the elasticity of substitution is a function of the relative factor combination v_2/v_1 only, independent of the scale of output $\bar{x}$, since x_1' and x_2' are homogeneous of degree zero and x_{12}'' is homogeneous of degree -1, cf. ALLEN, *op. cit.*, p. 319.

[3] Setting s_{12}, as evaluated in (9), equal to a (positive) constant s gives a second-order non-linear differential equation whose general solution for v_2 in terms of v_1 represents the general shape of an isoquant with constant elasticity of substitution. The solution is more easily obtained directly from the definition of s_{12}, setting $v_2/v_1 = u$ and $-dv_2/dv_1 = v$ and recalling that $du/u = d \log_e u$, $dv/v = d \log_e v$ so that we have

$$\frac{d \log_e u}{d \log_e v} = s \text{ (constant)}$$

which, by integration, gives

$$\log_e u = s \cdot \log_e v + K \text{ or } u = k \cdot v^s$$

where $k\,(= e^K > 0)$ is an arbitrary constant. Thus we have the first-order differential equation

(i) $$v_2/v_1 = k \cdot (-dv_2/dv_1)^s$$

or, separating the variables,

$$v_2^{-1/s} dv_2 = k_1 \cdot v_1^{-1/s} dv_1 \text{ (where } k_1 = -k^{-1/s} < 0)$$

whose integral for $s \neq 1$ is

(ii) $$v_2^{(s-1)/s} = k_1 \cdot v_1^{(s-1)/s} + k_2$$

where $k_1\,(<0)$ and k_2 are arbitrary constants. (For $s \to \infty$, the isoquant (ii) approaches a linear shape; for $s \to 0$ the differential equation (i) tends towards $v_2/v_1 = k$). For $s = 1$ the solution is

is a measure of the degree of substitutability along the isoquant in the neighbourhood of the particular point at which it is taken. Comparing two isoquants going through the same point and having a common tangent, the elasticity of substitution at that point will be greater for the flatter curve since it is inversely proportional to the curvature. If the shape of the isoquant approaches that of a straight line so that the two inputs become nearly perfect substitutes, s_{12} tends to infinity, whereas the elasticity tends to zero if the range of substitution for given output shrinks to a point so that the curvature tends to infinity[1].

For the general case of m inputs, $m > 2$, partial elasticities of substitution s_{ij} may be defined as between any pair of inputs (v_i, v_j).[2] These elasticities, however, are difficult to interpret in simple terms.

B. Production Curves and Diminishing Returns

(a) The isoquant diagram (Fig. 14) is one way of describing the production function, emphasizing the substitution aspect. Another is to study the behaviour of output for partial variation in one of the inputs (or for proportional variation in some inputs), all other inputs being kept constant. In the two-factor case, for constant $v_1 = \bar{v}_1$—e.g. for movement along $ABCD$ in Fig. 14—we have

$$x = x(\bar{v}_1, v_2)$$

whose graphical picture is a production curve (for parametric $\bar{v}_1$, a family of such curves).

The properties of the production curves can be deduced from those of the isoquant map. Thus, when the boundaries of the region of substitution are within the positive factor region—i.e., when there are definite limits to the amount of output that can be produced by increasing either input separately, the other input being kept constant—it follows as a corollary that the production curves $x = x(v_1, \bar{v}_2)$ and $x = x(\bar{v}_1, v_2)$ will be characterized by *diminishing returns* in the general sense that the variable factor's average productivity as well as its marginal productivity will be decreasing at least in a

---

$$\log_e v_2 = -k^{-1} \cdot \log_e v_1 + c$$

or

(iii) $$v_2 = c_1 \cdot v_1{}^{c_2}$$

where $c_1 \; (= e^c > 0)$ and $c_2 \; (= -k^{-1} < 0)$ are arbitrary.

Thus any isoquant for which the elasticity of substitution is the same at all points can be written in the form (ii) or (iii). The isoquants of a Cobb-Douglas function, $x = k \cdot v_1{}^\alpha \cdot v_2{}^\beta$ (α and β positive), belong to type (iii) and therefore have unit elasticity (whether or not $\alpha + \beta = 1$).

[1] It would be incorrect, however, to say that s_{12} between two limitational inputs is actually zero. Once the isoquant has been reduced to a single point, v_1 and v_2 are no longer independent variables in the production model—unless they are defined as amounts available (model (1c) in Ch. III, A), in which case the function, like that corresponding to discontinuous substitution, does not possess continuous derivatives so that s_{12} is not defined.

[2] Cf. ALLEN (1938), pp. 504 f. and 512 f.

neighbourhood of maximum output[1]. In Fig. 14, the maximum value of x along $ABCD$ is attained at C where x_2' changes in sign from positive to negative through zero[2]. Thus x_2' is decreasing ($x_{22}'' < 0$) in an interval containing point C. Average returns to the variable factor, x/v_2, will be a decreasing function of v_2 at least from some point between A and C since

$$\frac{\delta(x/v_2)}{\delta v_2} = \frac{v_2 x_2' - x}{v_2^2} \tag{10}$$

is negative not only for $x_2' \leq 0$ but also for sufficiently small positive values of x_2'. Decreasing average productivity is clearly equivalent to diminishing returns in the sense that the partial elasticity of production

$$\varepsilon_2 = \frac{\delta x}{\delta v_2} \cdot \frac{v_2}{x} = \frac{v_2 x_2'}{x}$$

is less than one.

---

[1] Historically, the principle of substitution in the theory of production is descended from observations of diminishing returns in agriculture, with land as the fixed factor which puts a limit to output. The rigorous treatment of the law of diminishing return dates back to RICARDO and v. THÜNEN, who made it the basis of a theory of the distributive shares of broad factor categories. Later in the nineteenth century the factors came to be treated more symmetrically in economic theory—WICKSTEED (1894), for example, used a production function explicitly with continuous substitution between all factors—and the analytical apparatus was subsequently applied to production within the individual firm. For modern treatment of production with continuous substitution see, e.g., FRISCH (1956), SCHNEIDER (1934) and (1958), SAMUELSON (1947), HICKS (1946), and CARLSON (1939).

A practical example of continuous factor substitution is given by SCHNEIDER (1958), pp. 170 and 189—193. A product is manufactured on a machine (e.g. a lathe) driven by a variable-speed motor and attended by a single operator. Motor fuel can be substituted for labour by speeding up the machine so that more fuel but fewer man-hours are required for a given quantity of product. The basic technological relation underlying the production function is that which gives the speed of the engine, z, as a function of fuel input per hour, u:

(i) $\qquad\qquad\qquad\qquad z = \varphi(u).$

Now if the hourly rate of production is proportional to the speed of the engine, the number of units produced per day will be

(ii) $\qquad\qquad\qquad x = c \cdot z \cdot v_1 \qquad (c = \text{constant})$

where v_1 is the number of hours the machine is operated per day, as measured in man-hours per day. Fuel consumption per day is

(iii) $\qquad\qquad\qquad\qquad v_2 = u \cdot v_1.$

Combining (i)–(iii) we get the production function

(iv) $\qquad\qquad\qquad x = c \cdot v_1 \cdot \varphi(v_2/v_1).$

Varying v_2 for v_1 constant, the resulting production curve will follow the shape of the function φ. The graphical picture of this function for a particular type of engine, based on engineering data—cf. op. cit., Fig. 84, p. 190—conforms to the familiar pattern of the law of variable proportions. (SCHNEIDER does not explicitly derive the production function (iv); using (i) and (iii) to express total cost per day in terms of the parameters u and v_1, he derives the cost function by minimizing cost subject to (ii) for parametric x.)—The model will have to be modified if the input of raw material is taken into account; see Ch. VII, B below.

Another empirical example of a production model with continuous substitution between two inputs is given by FRISCH (1935). However, this case is better analysed in terms of a three-factor model with two equations, cf. Ch. V, A.

[2] There will be no other local maxima when the isoquants are upward sloping outside the region of substitution: x_2' is positive at any point between A and B because $dv_2/dv_1 = -x_1'/x_2' > 0$ implies $x_2' > 0$ for $x_1' < 0$.

Similarly, partial variation of v_1 for v_2 constant will display diminishing returns to v_1 as x approaches its maximum with respect to v_1.

Whether or not the phase of diminishing average returns is preceded by a phase of increasing returns cannot be inferred without further assumptions. It is conceivable that x_2' and x/v_2 are decreasing functions of v_2 everywhere along $ABCD$, cf. *Fig. 16*.

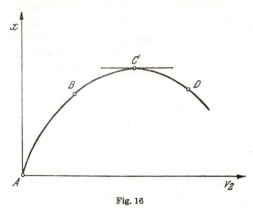

Fig. 16

Another possibility is that the function $x = x(\bar{v}_1, v_2)$ will have the specific shape of the classical *law of variable proportions*[1] as shown in *Fig. 17*.

A production curve of this kind is characterized by the following phases: The *marginal* productivity of the variable factor is positive up to point C, where $x_2' = 0$ and maximum output is attained, and negative beyond this point; in the region where it is positive, x_2' is first increasing with v_2 and then decreasing, having a maximum $(x_{22}'' = 0)$ at some point P between A and C. To the right of C, x_2' is still decreasing, at least in a neighbourhood of C; it may or may not have a minimum at some later point. *Average productivity* x/v_2 will be increasing through an initial phase $(x_2' > x/v_2,$

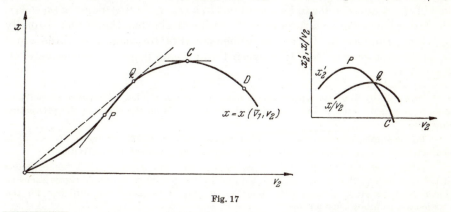

Fig. 17

---

[1] Also known as the "*law of diminishing returns*". This term will not be used here because it might be confused with the more general law of diminishing return stated above (i.e., decreasing x_2' and x/v_2 as the maximum limit to output—due to the constancy of the other factor—is approached).

$\varepsilon_2 > 1$), attaining a maximum somewhere between P and C (at point Q where $x_2' = x/v_2$ and $\varepsilon_2 = 1$); beyond Q there will be diminishing average returns ($0 < \varepsilon_2 < 1$ between Q and C, $\varepsilon_2 = 0$ at C where $x_2' = 0$, and $\varepsilon_2 < 0$ to the right of C).

The marginal and average productivity curves are also shown in Fig. 17, where x_2' is seen to intersect x/v_2 from above at the latter's maximum point.

In the important case where the production function $x = x(v_1, v_2)$ is homogeneous of the first degree, it is possible to derive production curves conforming to this pattern provided that the function satisfies a few additional simple assumptions[1]. However, homogeneity is neither necessary nor sufficient for the law of variable proportions to hold for partial variation in one of the inputs[2].

(b) More general cases where the number of inputs exceeds two can be examined in a similar way for partial variation of a single input. When two or more inputs are allowed to vary simultaneously, the others being constant, the analysis gets more complicated because the variation of x can no longer be studied along a two-dimensional production curve.

However, as long as we consider *proportional* variation of the inputs selected as variables, the production model reduces to a two-factor case and can be treated as such. For example, the production function

$$x = x(v_1, v_2, v_3)$$

can be written

$$x = \Phi(v_1, v_2)$$

for variations such that $v_3 = \beta v_2$ where β is some constant. For constant $v_1 = \bar{v}_1$, it seems plausible—though by no means certain—that x will respond to variations in v_2 in a manner which preserves at least some of the characteristic features of the law of variable proportions, in particular the typical behaviour of the elasticity $\varepsilon_2 = (dx/dv_2) \cdot v_2/x = \Phi_2' \cdot v_2/x$.[3] If this is so for any value of β (within the relevant range of values of this parameter), the loci of $\varepsilon_2 = 1$ and $\varepsilon_2 = 0$ can be traced out in a (v_2, v_3) factor diagram ($v_1 = \bar{v}_1$); the latter locus will be farther away from the origin than the former. Broadly speaking, the loci divide the two-dimensional factor plane into three zones in which ε_2 is greater than 1, between 1 and 0, and negative respectively, cf. *Fig. 18.* [4]

---

[1] In Appendix 2 an attempt has been made to deduce the law of variable proportions from homogeneity with a minimum of further restrictions on the shape of the production function.

[2] For example, a homogeneous Cobb-Douglas function $x = k \cdot v_1^a \cdot v_2^{1-a}$ ($0 < \alpha < 1$) is characterized by monotonically decreasing marginal and average productivities.

[3] If the production function $x = x(v_1, v_2, v_3)$, and thus also $x = \Phi(v_1, v_2)$, is homogeneous of the first degree, this pattern can be deduced from a minimum of simple assumptions as shown in Appendix 2. Homogeneity, however, is neither necessary nor sufficient for the elasticity ε_2 to follow such a course.

[4] The definition of ε_2 is not dependent on a linear expansion path through the origin: at any point (v_2^0, v_3^0), ε_2 is defined as the elasticity of x with respect to small proportional variations (dv_2, dv_3) where $dv_3 = (v_3^0/v_2^0) \cdot dv_2$. Its value at the point does not depend on what particular expansion path through the point is considered.

This suggests that ε_2 will vary in a similar manner along any increasing (or at least non-decreasing) expansion path as shown in Fig. 18—the more so, the nearer the variation is to proportionality between v_2 and v_3. A similar argument applies to the general case of m inputs some of which are kept con-

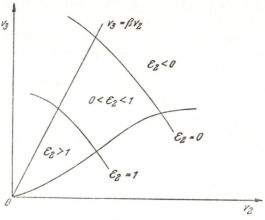

Fig. 18

stant. A phase of increasing returns to the variable factor complex ($\varepsilon_2 > 1$) will be followed by a phase of diminishing returns ($\varepsilon_2 < 1$) and eventually also by decreasing output ($\varepsilon_2 < 0$), just as in the two-factor case[1]. The similarity to the law of variable proportions is even more marked if ε_2 is assumed to decrease monotonically along any linear expansion path (for constant β) and, therefore, along any positively inclined path because all loci of constant ε_2 are intersected in the same order.

These results—which are not restricted to three-factor cases with two variable inputs—have immediate bearing on cost behaviour. With a production function in m inputs of which M are fixed factors, the (short-run) cost function is defined along an expansion path in $(m - M)$-dimensional factor space. The elasticity of output with respect to the variable factor complex turns out to be the reciprocal elasticity of cost with respect to output[2] and, therefore, describes the shape of the cost function in the neighbourhood of the point at which it is taken. Hence, provided that the elasticity behaves in the way indicated above, the cost function will have roughly the same shape as in the two-factor case, that is, the shape of the inverse of the productivity curve $x = x(\bar{v}_1, v_2)$.[3]

C. An Alternative Model

It might be argued against the above analysis that points at which one or more marginal productivities is less than zero should not be included in the production function since they involve an obvious waste. In *Fig. 19*,

---

[1] Cf. SCHNEIDER (1934), p. 18.

[2] Cf. IV, E below.

[3] This result requires modification, however, when the capacity limits are of the type $v_s \leq \bar{v}_s$ rather than $v_s = \bar{v}_s$; cf. Ch. IV, E below and Ch. VII.

P on the isoquant $x(v_1, v_2) = \bar{x}$ is such an inefficient point, x_2' being negative. Clearly a larger output can be obtained by leaving some of the second input idle, using only the combination represented by point Q, where $x(v_1, v_2) = \bar{\bar{x}} > \bar{x}$. However, this is merely a matter of whether to define the v_i as quantities actually used—in this case, the factor combination P results in output $x = \bar{x}$—or to define them as available quantities, allowing for less than complete utilization of the amounts that happen to be at the firm's disposal. In the latter case, when P represents the factor combination available, any amount of product from 0 to $\bar{\bar{x}}$ may be forthcoming, depending on the extent to which the factors are utilized, since any isoquant for which $0 \leq x \leq \bar{\bar{x}}$ has points in common with the availability rectangle $OAPB$. In order to make the production function single-valued, then, it must be defined as giving the largest output that can be made with given input amounts available, in this case $\bar{\bar{x}}$ with the factor combination P.[1]

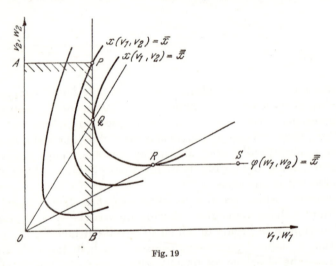

Fig. 19

Analytically, if (w_1, w_2) are inputs available and (v_1, v_2) inputs consumed, this means defining the production function $x = \varphi(w_1, w_2)$ as

$$x = \text{Max } x(v_1, v_2) \text{ for } v_1 \leq w_1, v_2 \leq w_2. \quad [2] \tag{11}$$

Within the region of substitution, this makes no difference to the shape of the isoquant: all points on QR in Fig. 19 represent the same output $\bar{x}$ no matter how the inputs are defined. Outside the region, the positively inclined branches of the isoquants are replaced by straight lines parallel to the axes,

---

[1] Many authors define the production function in this way, for example, SAMUELSON (1947), p. 58, or CARLSON (1939), pp. 14 f. and 26.

[2] The maximization involved in this definition goes further than that underlying the production function (1), where it was merely assumed that the factors are always organized in a technically optimal way for any given set of factor amounts $v_1, v_2, \ldots, v_m$, whether or not output could be increased by using some other factor combination with less of one or more factors and the same amounts of the others. Maximal x must be interpreted differently according as the inputs are defined as v_i or w_i, cf. point P in Fig. 19.

as exemplified by $PQRS$ (for $x=\bar{\bar{x}}$) [1]—just as the isoquants in the limiting case of limitational inputs will be L-shaped instead of points when the inputs are defined as available quantities[2].

This, of course, has important consequences as to the shape of the production curves, cf. *Fig. 20a–b*. For a given amount available of the first input, $w_1 = \bar{w}_1$, the curve $x = \varphi(\bar{w}_1, w_2)$, as defined along $ABCD$ in Fig. 20a, will have the shape of Fig. 20b (though the phase APB will be linear only when $x = x(v_1, v_2)$ is homogeneous of the first degree).

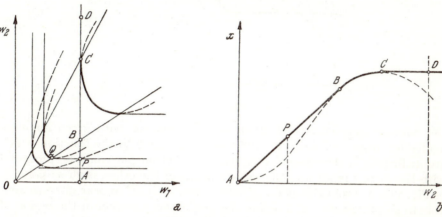

Fig. 20a-b

Along AB in Fig. 20a, where the first input would have a negative marginal productivity if used to the full, a larger output will be forthcoming for any $w_2 = v_2$ if some of $\bar{w}_1$ is left idle (cf. points P and Q); the factor combination actually used (v_1, v_2) will follow the path OB (which in the case of homogeneity is linear and characterized by constant (positive) marginal productivity with respect to the second factor)[3].

Along BC, i.e., in the region of substitution, the curve will coincide with $x = x(\bar{v}_1, v_2)$ since both inputs will be used up ($\bar{v} = \bar{w}_1$, $v_2 = w_2$). Beyond the maximum point C, however, the production curve will be horizontal[4] since the expansion path coincides with the vertical branch of the isoquant: at D, the combination (v_1, v_2) used, and thus also the output, will be the same as in C, the excess amount of w_2 being left idle because its application would be harmful. Thus, φ_2' will be zero beyond C. (This also applies to a non-homogeneous production function.) The marginal productivity pattern $ABCD$ corresponding to Fig. 20b is shown in *Fig. 21*.

---

[1] CARLSON (1939), p. 26, draws the isoquant map in this way.

The relationship between the two different isoquant patterns and the underlying assumptions was first analysed by GLOERFELT-TARP (1937), pp. 258 f.

[2] Cf. also the case of discontinuous substitution (Ch. III, B above).

[3] Cf. GLOERFELT-TARP, *op. cit.*, p. 233, who also observed (p. 254) that any point on AB in Fig. 20a can be represented as what was later to be known as a convex combination of two linear activities, OB and OA, of which the latter is a slack activity which produces nothing but makes away with the idle part of w_1. In linear programming terms this was later shown by CHIPMAN (1953), p. 106, with reference to the production curve (Fig. 20b).

[4] Cf. GLOERFELT-TARP (1937), p. 256, or CHIPMAN (1953), p. 106.

This way of defining the production function has the advantage of eliminating one obvious source of inefficiency, the waste caused by negative marginal productivities. For any combination (w_1, w_2) we have $\varphi_1' \geq 0$, $\varphi_2' \geq 0$. The production function $x = \varphi(w_1, w_2)$ is useful in that it answers

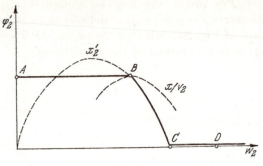

Fig. 21

the relevant question: How much can be produced with given amounts available? However, there are still points belonging to the function which are inefficient in the sense that a given level of output could be produced with less of one input and the same amount of the other, such as point D in Fig. 20a–b. From this point of view, then, it would appear more reasonable to go all the way and define the production function in the region of substitution only. Whichever definition is adopted, the optimum point—minimum cost or maximum profit—will always be some point in this region, where the two functions $x = x(v_1, v_2)$ and $x = \varphi(w_1, w_2)$ coincide, but the former kind of production function—even when defined over the entire non-negative region—has the distinct advantage of being analytically more convenient for purposes of optimization.

D. Cost Minimization

1. Marginal Equilibrium Conditions

(a) The production function serves as a side condition subject to which cost is to be minimized or—more generally—profit is to be maximized.

To find the *least-cost point* on the isoquant $x = \bar{x}$, minimize

$$c = \sum_{i=1}^{m} q_i v_i$$

subject to

$$x(v_1, v_2, \ldots, v_m) = \bar{x}, \tag{12}$$

i.e., minimize the Lagrangian[1]

$$L = \sum_{i=1}^{m} q_i v_i + u \cdot [\bar{x} - x(v_1, v_2, \ldots, v_m)]$$

where u is a Lagrange multiplier, a provisionally undetermined constant.

[1] Cf. Appendix 1, A.

The *necessary* first-order conditions for a minimum are

$$dL = \sum_{i=1}^{m} \frac{\delta L}{\delta v_i} dv_i = 0$$

or

$$\frac{\delta L}{\delta v_i} = q_i - u \cdot x_i' = 0 \qquad (i = 1, 2, \ldots, m), \qquad (13)$$

which, with (12), determine u and the v_i.[1] Geometrically, this means that the isocost plane through the point is tangent to the isoquant surface: at the common point, the q_i will be proportional to the x_i' so that the gradient of the isocost plane in any direction, $\delta v_m / \delta v_i = -q_i/q_m$ $(i = 1, 2, \ldots, m-1)$, is equal to that of the tangent plane, $\delta v_m / \delta v_i = -x_i'/x_m'$.

Assuming that there exists one or more non-negative (and thus feasible) solutions to (12)–(13)[2], solutions involving negative marginal productivities can be discarded at once since an inefficient point can never be optimal[3]; in other words, only points in the region of substitution need be considered.

To ensure that a feasible, efficient point satisfying (12)–(13) represents a minimum rather than a maximum, additional information is required. If the nature of the solution is not obvious—e. g. from geometrical considerations in the two-factor case—the second-order condition for a constrained minimum provides a criterion. It is *sufficient* for the solution of (12)–(13) to represent a (local) minimum that

$$d^2 L = \sum_{i=1}^{m} \sum_{j=1}^{m} \frac{\delta^2 L}{\delta v_i \delta v_j} dv_i dv_j > 0$$

at the point for all variations satisfying (12), i.e., subject to

$$\sum_{i=1}^{m} x_i' dv_i = 0.$$

Since, for constant prices, $\dfrac{\delta^2 L}{\delta v_i \delta v_j} = -u \cdot x_{ij}''$ where $u = q_i/x_i' > 0$ at an efficient point, this boils down to

$$\sum_{i=1}^{m} \sum_{j=1}^{m} x_{ij}'' dv_i dv_j < 0 \quad \text{for} \quad \sum_{i=1}^{m} x_i' dv_i = 0. \qquad (14)$$

---

[1] WALRAS, of all people, was the first to give (in the later editions of his *Éléments*) a rigorous treatment of the conditions for minimum cost under continuous substitution; cf. WALRAS (English ed., 1954), pp. 384 f., 549 f., and 604 f.

[2] If the isoquant is linear, the least-cost point is generally not characterized by (13); cf. below.

[3] Cf. Ch. II above. If $x_i' < 0$ for some i, a marginal decrease in the corresponding v_i will raise output; if accompanied by appropriate marginal decreases in those inputs for which $x_i' > 0$, we can get back to some point on the isoquant $x = \bar{x}$, i.e., there exist factor combinations which produce this amount using less of all factors (except those for which $x_i' = 0$, which are left as they were) and thus at less cost.

The quadratic form in (14) will be negative definite subject to the linear constraint $\sum_i x_i' dv_i = 0$ if and only if

$$(-1)^k \begin{vmatrix} 0 & x_1' & \ldots & x_k' \\ x_1' & x_{11}'' & \ldots & x_{1k}'' \\ \cdots\cdots\cdots\cdots\cdots \\ x_k' & x_{k1}'' & \ldots & x_{kk}'' \end{vmatrix} > 0 \qquad (k = 2, 3, \ldots, m)\,^1 \qquad (15)$$

which is therefore an equivalent formulation of the sufficient condition (14).

Condition (14), and thus also (15), is automatically satisfied if the isoquant is strictly convex to the origin in the region of substitution, cf. (7). Writing the equation of the isoquant in the form $v_m = f(v_1, v_2, \ldots, v_{m-1})$ and expanding by Taylor's series[2] with a remainder R_1, we get

$$f(v_1, \ldots, v_{m-1}) = f(v_1^0, \ldots, v_{m-1}^0) + \sum_{i=1}^{m-1} [f_i']^0 (v_i - v_i^0) + R_1$$

where

$$R_1 = \frac{1}{2} \sum_{i=1}^{m-1} \sum_{j=1}^{m-1} f_{ij}'' \cdot (v_i - v_i^0)(v_j - v_j^0),$$

f_{ij}'' being taken at some intermediate point of the isoquant. By (7), R_1 is positive when the isoquant is convex; assuming f_{ij}'' to be continuous, the quadratic form will have the same sign if f_{ij}'' is taken at $(v_1^0, \ldots, v_{m-1}^0)$ for sufficiently small $v_i - v_i^0$ and $v_j - v_j^0$. Hence, the fixed point being arbitrary, we have at any efficient point of the isoquant

$$\frac{1}{2} \sum_{i=1}^{m-1} \sum_{j=1}^{m-1} f_{ij}'' dv_i dv_j > 0. \qquad (16)$$

For infinitesimal movements along the isoquant from the point we have

$$dx = \sum_{i=1}^{m} x_i' dv_i = 0$$

and, with v_m as the dependent variable,

$$d^2x = d(dx) = \sum_{i=1}^{m} \sum_{j=1}^{m} x_{ij}'' dv_i dv_j + x_m' d^2v_m = 0.$$

Since, by (16),

$$d^2v_m = \sum_{i=1}^{m-1} \sum_{j=1}^{m-1} f_{ij}'' dv_i dv_j > 0$$

and $x_m' > 0$ in the interior of the region of substitution, convexity is seen to imply that (14)—or the equivalent condition (15)—is satisfied everywhere in the region[3]. For $m = 2$, (16) becomes

---

[1] Cf. Appendix 1, A.

[2] Cf. ALLEN (1938), pp. 456 ff.

[3] Convexity is sometimes defined by (15): the isoquant is convex to the origin if (15) is satisfied at any point in the interior of the region of substitution. Cf., e.g., FRISCH (1956), pp. 80 and 147.

$$\tfrac{1}{2} f_{11}'' \, dv_1^2 > 0 \quad \text{or} \quad f_{11}'' = d^2 v_2 / dv_1^2 > 0,$$

cf. (6).

Moreover, convexity throughout the region of substitution guarantees that the solution determined by (12)–(13) is unique. At a feasible, efficient point $(v_1^0, \ldots, v_m^0)$ satisfying (13) we have

$$(x_i')^0 = q_i / u$$

which with (8) gives

$$\frac{1}{u} \cdot \sum_{i=1}^{m} q_i \cdot (v_i - v_i^0) > 0$$

for all points $(v_1, \ldots, v_m)$ on the isoquant (not all $v_i = v_i^0$).[1] Since u is positive for some or all q_i positive and $x_i' \geq 0$, this implies

$$\sum_{i=1}^{m} q_i v_i^0 < \sum_{i=1}^{m} q_i v_i,$$

that is, an efficient point which satisfies the necessary conditions is cheaper than any other factor combination. Thus, for a convex isoquant, (13) is necessary as well as sufficient for minimum cost and the solution represents a *global* minimum. (In the two-factor case this is geometrically obvious.)

Had the isoquant been alternately convex and concave in the region of substitution, there would have been several points satisfying (13), some of which represent local minima—such as P and R in *Fig. 22*, of which R is the global minimum at the particular set of prices—whereas others such as Q and S, where the isoquant is concave, are local maxima[2].

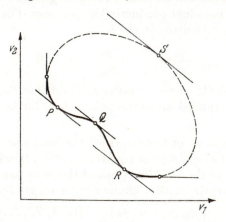

Fig. 22

---

[1] This holds whether $(v_1, \ldots, v_m)$ is efficient or not, cf. above.

[2] Fig. 22 shows that, whereas point S can be discarded straight away as being inefficient $(x_1' < 0, x_2' < 0)$, it is possible for a point such as Q to be efficient although it can never represent a least-cost solution for any set of prices; Q cannot be ruled out by the technological efficiency criterion when x_1' and x_2' are positive at the point—unless the "bulge" can be replaced by a straight line representing combinations of points on the convex parts of the isoquant (cf. Ch. IV, A above).

When the factor prices q_i are all positive, it follows from (13) that $x_i' > 0$ for all i so that the least-cost factor combination will be a point in the interior of the region of substitution such as P in *Fig. 23*. (13) can then be written in the form

$$(u=) \frac{q_1}{x_1'} = \frac{q_2}{x_2'} = \cdots = \frac{q_m}{x_m'},$$

the interpretation of which is that the marginal cost for partial factor variation[1], $dc/dx = q_i dv_i/dx = q_i/x_i'$ (all other inputs constant), will be the same in all directions.

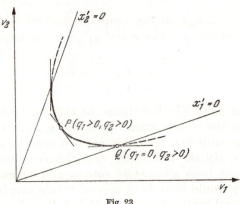

Fig. 23

On the other hand, if one or more factor prices is zero, the optimal solution will be a point on the boundary of the region of substitution because the corresponding marginal productivities are zero. For, say, $q_1 = 0$ conditions (13) can then be written in the form

$$\frac{q_2}{x_2'} = \cdots = \frac{q_m}{x_m'}, \quad x_1' = 0$$

which together with (12) determine the v_i. Point Q in Fig. 23 is an example: the free input v_1 is applied up to the point where further application would be inefficient[2].

(b) For any given set of factor prices, the least-cost factor combination is determined by the equilibrium equations (12)–(13), which implicitly give the optimal values of the v_i as functions of the price parameters $q_1, \ldots, q_m$. Solving (12)–(13) for the v_i after eliminating u we get the explicit functions

$$v_i = v_i(q_1, q_2, \ldots, q_m) \qquad (i = 1, 2, \ldots, m) \tag{17}$$

which, by the assumptions made above, are single-valued, continuous and

---

[1] In FRISCH's terminology, the "partial marginal cost in the direction of v_i", cf. FRISCH (1956), p. 146.

[2] When an input is free, this is usually because it represents the services of a fixed factor of limited capacity. If the solution (point Q) does not respect the corresponding capacity limitation $v_1 \leq \bar{v}_1$—i.e., if the line $v_1 = \bar{v}_1$ intersects the isoquant somewhere in the region of substitution—it is not feasible and the problem will have to be reconsidered with the capacity restriction as an additional side condition; see below.

differentiable so that the least-cost combination will respond in a continuous manner to factor price changes. Furthermore, the functions are homogeneous of degree zero: proportional variation in all q_i leaves the solution for the v_i unaltered.

The displacement of the equilibrium position brought about by given *finite* changes in one or more factor prices is best studied by comparing alternative solutions to (12)–(13). The effect of a *marginal* change in a single q_i, on the other hand, can be traced by differentiating the system (12)–(13) with respect to the price in question. In the two-factor case write the system in the form

$$x(v_1,\, v_2) = \bar{x} \tag{12a}$$

$$q_1 \cdot x_2'(v_1,\, v_2) - q_2 \cdot x_1'(v_1,\, v_2) = 0\,; \tag{13a}$$

differentiating with respect to, say, q_2 (q_1 constant) we get

$$x_1' \cdot \frac{\delta v_1}{\delta q_2} + x_2' \cdot \frac{\delta v_2}{\delta q_2} = 0$$

$$(q_1 x_{12}'' - q_2 x_{11}'') \cdot \frac{\delta v_1}{\delta q_2} + (q_1 x_{22}'' - q_2 x_{12}'') \cdot \frac{\delta v_2}{\delta q_2} = x_1'$$

which is a linear equation system in the partial derivatives with respect to q_2 of the functions $v_1 = v_1(q_1, q_2)$, $v_2 = v_2(q_1, q_2)$. (The coefficients are constant because the derivatives of the production function are taken at the particular point of equilibrium.)

Solving the equations, and making use of (13a) to eliminate q_1 in the determinant of the system, we have

$$\frac{\delta v_1}{\delta q_2} = \frac{-x_1'(x_2')^2}{q_2 D}, \quad \frac{\delta v_2}{\delta q_2} = \frac{(x_1')^2 x_2'}{q_2 D} \tag{18}$$

where

$$D = (x_2')^2 x_{11}'' - 2 x_1' x_2' x_{12}'' + (x_1')^2 x_{22}''.$$

In elasticity form we have

$$\frac{\delta v_1}{\delta q_2} \cdot \frac{q_2}{v_1} = \frac{-x_1'(x_2')^2}{v_1 D}, \quad \frac{\delta v_2}{\delta q_2} \cdot \frac{q_2}{v_2} = \frac{(x_1')^2 x_2'}{v_2 D}. \tag{19}$$

These partial elasticities, which are independent of the units of measurement, are functions of v_1 and v_2; their values depend on the properties of the production function—the first and second derivatives—at the particular equilibrium point at which they are taken. Their respective signs, however, will be the same at any point. By the assumption of convexity (6) we have $D < 0$ [1] at a least-cost point in the interior of the region of substitution where x_1' and x_2' are both positive. It follows that

$$\frac{\delta v_1}{\delta q_2} > 0, \quad \frac{\delta v_2}{\delta q_2} < 0$$

---

[1] This result also follows from the second-order minimum conditions (15): for $k = 2$, the determinant which is to be positive is equal to $-D$.

and the elasticities will also be positive and negative respectively—that is, a fall in q_2 will lead to a substitution of v_2 for v_1, as is also geometrically obvious.

In the general m-factor case it can similarly be shown that more will be used of an input which has become cheaper $(\delta v_i / \delta q_i < 0)$.[1] The cross derivatives $\delta v_j / \delta q_i$ $(i \neq j)$ may vary in sign. Differentiating the general equilibrium equations

$$x(v_1, v_2, \ldots, v_m) = \bar{x} \tag{12}$$

$$u \cdot x_i' - q_i \quad\quad = 0 \quad\quad (i = 1, 2, \ldots, m) \tag{13}$$

with respect to, for example, q_m, we have

$$\sum_{j=1}^{m} x_j' \cdot \frac{\delta v_j}{\delta q_m} = 0$$

$$x_i' \cdot \frac{\delta u}{\delta q_m} + u \cdot \sum_{j=1}^{m} x_{ij}'' \cdot \frac{\delta v_j}{\delta q_m} = 0 \quad\quad (i = 1, 2, \ldots, m-1)$$

$$x_m' \cdot \frac{\delta u}{\delta q_m} + u \cdot \sum_{j=1}^{m} x_{mj}'' \cdot \frac{\delta v_j}{\delta q_m} = 1 .$$

After dividing through by u in all but the first equation the system can be solved for $\delta v_m / \delta q_m$ to give

$$\frac{\delta v_m}{\delta q_m} = \frac{(1/u^2) \cdot \Delta_{mm}}{(1/u) \cdot \Delta}$$

where Δ is the bordered determinant

$$\Delta = \begin{vmatrix} 0 & x_1' & \ldots & x_m' \\ x_1' & x_{11}'' & \ldots & x_{1m}'' \\ \cdot & \cdot & & \cdot \\ \cdot & \cdot & & \cdot \\ \cdot & \cdot & & \cdot \\ x_m' & x_{m1}'' & \ldots & x_{mm}'' \end{vmatrix}$$

and Δ_{mm} is the determinant obtained from Δ by deleting the last row and the last column. The second-order minimum conditions as written in the form (15) imply that Δ and Δ_{mm} are of opposite signs, and u is positive; hence

$$\frac{\delta v_m}{\delta q_m} < 0, \text{ i.e., } \frac{\delta v_i}{\delta q_i} < 0 \text{ for all } i$$

as the numbering is immaterial.

The cross derivatives $\delta v_j / \delta q_i$ cannot be shown to be all positive as in the two-factor case, though the positive signs can be expected to be dominant since more of v_i must be offset by less of some or all other inputs if x is to remain constant. Nor can a definite conclusion with respect to the sign of $\delta u / \delta q_m$ be reached by solving the system for this variable.

---

[1] Cf. SAMUELSON (1947), pp. 63 ff.

In the two-factor case, the displacement of the least-cost point caused by a marginal change can also be expressed by means of the elasticity of substitution s_{12}, as defined by (9). The optimum point is characterized by

$$\frac{q_1}{q_2} = -\frac{dv_2}{dv_1}\left(=\frac{x_1{}'}{x_2{}'}\right);$$

hence s_{12} also expresses the elasticity of the relative factor combination v_2/v_1 with respect to marginal changes in the price ratio q_1/q_2. Clearly this is a more convenient way of measuring the substitution involved in shifting the optimum position: all the relevant information conveyed by the derivatives $\delta v_i/\delta q_j$ ($i=1, 2; j=1, 2$) has been compressed into a single number which is independent of the units of measurement—though of course not independent of the particular optimum point at which it is taken[1].

Thus the displacement of the relative factor combination for a given change in the relative prices is greater, the greater the elasticity of substitution is—i.e., the flatter the isoquant is—at the initial optimum point. This is illustrated by the two isoquants in *Fig. 24a*, where point P is the original optimum whereas Q and R represent the least-cost points after the price change.

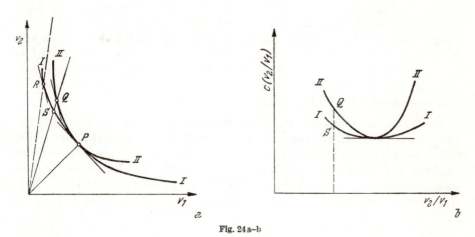

Fig. 24a–b

---

[1] The elasticities of v_1 and v_2 with respect to the factor prices are related in a simple way to the elasticity of substitution. Evaluating s_{12} in terms of the derivatives of the production function we get (cf. Ch. IV, A above)

$$s_{12} = \frac{-x_1{}'x_2{}'(v_1x_1{}'+v_2x_2{}')}{v_1v_2D}$$

which, with (19) and (13), leads to

$$\frac{\delta v_1}{\delta q_2}\cdot\frac{q_2}{v_1} = \frac{v_2x_2{}'}{v_1x_1{}'+v_2x_2{}'}\cdot s_{12} = \frac{q_2v_2}{q_1v_1+q_2v_2}\cdot s_{12}$$

and

$$\frac{\delta v_2}{\delta q_2}\cdot\frac{q_2}{v_2} = \frac{-v_1x_1{}'}{v_1x_1{}'+v_2x_2{}'}\cdot s_{12} = \frac{-q_1v_1}{q_1v_1+q_2v_2}\cdot s_{12}.$$

The numerical values of the two elasticities add up to s_{12}.

The elasticity of substitution—or the curvature of the isoquant—also provides a measure of the "flatness" of the cost minimum, in the sense that the increase in total cost resulting from a given (positive or negative) deviation from that relative factor combination which is cheapest at the given prices will be less, the greater the elasticity of substitution is at the least-cost point. This is illustrated in *Fig. 24b*, where the curves indicate the cost of producing the given amount of output as a function of the relative factor combination applied, the minimum point corresponding to point P in Fig. 24a. Curve I, which corresponds to the isoquant I in Fig. 24a, represents the flatter minimum[1].

2. Boundary Minima

(a) The necessary conditions (13) were derived by minimizing cost subject only to the production function for constant x, disregarding such additional constraints on the variables as non-negativity requirements $(v_i \geq 0)$ and possibly also capacity restrictions $(v_s \leq \bar{v}_s)$.

These boundary conditions, which set limits to the region of feasible solutions, present no problems if the solution to (12)–(13) happens to respect them. But if it turns out to violate one or more of the inequalities, the solution is infeasible and therefore cannot represent an optimum, being devoid of economic meaning. It may also happen that no solution to (12)–(13) exists. In either case the least-cost solution will be a point on the boundary of the feasible region where one or more of the boundary conditions is satisfied in equality form. To find such a boundary solution other methods of constrained minimization must be resorted to, which take explicit account of side conditions in inequality form[2].

(b) When the isoquant is convex in the region of substitution so that the marginal rates of substitution vary continuously from zero to infinity, there will always exist a solution to (12)–(13) which satisfies the second-order condition (14). However, this does not apply to the limiting case of perfect substitutability where the isoquants are linear and thus not strictly convex. (12) will have the form

$$\sum_{i=1}^{m} b_i v_i = a \tag{20}$$

where a, and possibly also the coefficients b_i, will depend on the output

---

[1] The importance of examining the "flatness" of an economic optimum—in the present case, the curvature of the function $c(v_2/v_1)$ for given x in the neighbourhood of the minimum—has been pointed out by FORCHHAMMER (1937), pp. 114 ff. This point has generally been neglected in economic theory, presumably because of the absence of empirical data. The consequences of deviating from the optimum point, as expressed for example in a graph like Fig. 24b, are more likely to interest engineers and production managers.

[2] Cf. Appendix 1, B–C. The modified Lagrange method, due to KUHN and TUCKER, provides a general criterion for maximum or minimum of a function subject to constraints some or all of which have the form of inequalities. Many simple cases, however, may be attacked by more straightforward methods as shown in the following. Linear problems are best solved by linear programming methods such as the simplex method.

parameter $\bar{x}$.[1] Clearly in this case the least-cost point cannot in general be characterized by proportionality between factor prices and marginal productivities; a finite minimum of cost exists only because the v_i cannot be negative, and the least-cost point will be a boundary solution, determined by minimizing c subject to (20) and to $v_i \geq 0$. This is a simple linear programming problem.

In the two-factor case of *Fig. 25*, the solution is obviously either point $A = (0,\, a/b_2)$ or $B = (a/b_1,\, 0)$, according as to whether the slope of the iso-cost lines, $-q_1/q_2$, is less than or greater than the constant slope of the isoquant AB, $-b_1/b_2$, unless they happen to be equal so that any point on AB—including the extreme points—will be optimal.

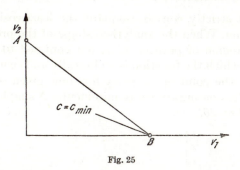

Fig. 25

In the general case of m inputs there will always exist a least-cost point at which only one of the v_i is positive, the other inputs not being used[2]. Solving (20) for, say, v_1 as the basic variable and substituting in the cost expression we have

$$v_1 = \frac{a}{b_1} - \sum_{i=2}^{m} \frac{b_i v_i}{b_1} \quad \text{and} \quad c = \frac{q_1 a}{b_1} + \sum_{i=2}^{m} \left(q_i - \frac{q_1 b_i}{b_1} \right) v_i.$$

Now suppose that

---

[1] Any production function which can be written in the form

$$\sum_{i=1}^{m} f_i(x) \cdot v_i = F(x)$$

will have isoquants of the form (20), cf. above. The special case of the linear and homogeneous function

$$x = \sum_{i=1}^{m} b_i v_i$$

where the b_i are constants (to be identified with the $x_i{}'$) may be thought of as a linear discontinuous-substitution model with m activities:

$$x = \sum_{k=1}^{m} \lambda_k, \quad v_i = \sum_{k=1}^{m} a_{ik} \lambda_k \quad \text{where} \quad a_{ik} = \begin{cases} 1/b_i & \text{for } k = i \\ 0 & \text{for } k \neq i, \end{cases}$$

i.e., each activity produces the output using only one input. Geometrically this means that the activity vectors are the respective axes of the factor diagram.

[2] This follows from the fundamental theorem of linear programming; apart from the non-negativity requirements there is only one side condition, (20).

$$q_i - \frac{q_1 b_i}{b_1} \geq 0 \quad \text{for } i = 2, 3, \ldots, m. \tag{21}$$

Then the simplex criterion, which is always a sufficient condition for the basic feasible solution to be optimal, is satisfied and c is a minimum for $v_1 = a/b_1$, $v_2 = v_3 = \cdots = v_m = 0$. [1] (The least-cost point is unique if (21) is satisfied in strict inequality form for $i = 2, 3, \ldots, m$).

The optimum point will be completely insensitive to price changes except at the critical price ratios where the response is discontinuous: a rise in q_1 will not affect the solution until one of the simplex coefficients in (21) gets down to zero and the optimum combination switches to a point where some other input is used exclusively[2].

(c) Even for a strictly convex isoquant the least-cost point may be a boundary minimum. When the analytical shape of the production function is such that the region of positive x_i' is not confined to the non-negative factor space over which the function is technically and economically defined, it is possible that the point of tangency for some given set of factor prices fails to meet all the non-negativity requirements. A simple case of this kind is illustrated by *Fig. 26*.

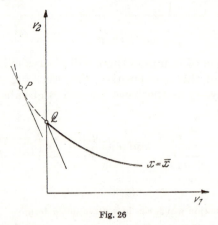

Fig. 26

The analytical solution to (12)–(13), as represented by point P, is not feasible since v_1 is negative. The best feasible solution is obviously point Q where $v_1 = 0$, i.e., a point on the boundary of the economic region of defini-

---

[1] If (20) has the form

$$\sum_i b_i v_i = \bar{x} \qquad (\text{i.e., } b_i = x_i'),$$

condition (21) means that

$$\frac{q_1}{x_1'} \leq \frac{q_i}{x_i'} \qquad (i = 2, 3, \ldots, m),$$

i.e., marginal cost is always least in the direction of v_1, which will therefore be substituted for all other inputs up to the point where v_2, v_3, etc. become zero.

[2] Thus, paradoxically, the optimum factor combination will show great stability with respect to price variations in the case of extreme substitutability, just as it does in the opposite extreme case of limitationality; cf. LEONTIEF (1951), pp. 40 f.

tion; this will be so for any pair of prices satisfying the inequality

$$(-\infty <) -\frac{q_1}{q_2} \leq -\frac{(x_1')^Q}{(x_2')^Q} \quad \text{or} \quad \frac{q_1}{(x_1')^Q} \geq \frac{q_2}{(x_2')^Q}$$

which replaces the corresponding equality derived from conditions (13). c is greater at Q than at P.

In the general case of m inputs, a similar expedient suggests itself when Lagrange's method fails to produce a feasible solution. Let $P = (v_1^P, v_2^P, \ldots, v_m^P)$ represent the minimum of c subject to (12), characterized by

$$q_i = u \cdot (x_i')^P \qquad (i = 1, 2, \ldots, m), \tag{22}$$

cf. (13). Now suppose that $v_1^P < 0$ but all other $v_i^P \geq 0$. Discarding the solution P as infeasible, let us try—as we did in the two-factor example—to set $v_1 = 0$ beforehand and see whether this leads to a feasible least-cost solution. Minimizing the cost expression subject to (12) with $v_1 = 0$—the equality form of the boundary condition $v_1 \geq 0$ which was violated by the solution P—we get a new solution $Q = (v_1^Q, v_2^Q, \ldots, v_m^Q)$, determined by the side conditions $x(v_1^Q, v_2^Q, \ldots, v_m^Q) = \bar{x}$, $v_1^Q = 0$, and the necessary minimum conditions

$$q_i = \lambda \cdot (x_i')^Q \qquad (i = 2, 3, \ldots, m) \tag{23}$$

where λ is a Lagrange multiplier[1]. It can be shown that this solution, if feasible and efficient ($v_i^Q \geq 0$, $(x_i')^Q \geq 0$), is cheaper than any other feasible point on the isoquant.

For all $q_i \geq 0$ we have $(x_i')^P \geq 0$ for all i. Assuming that Q is also a point in the region of substitution, i.e., $(x_i')^Q \geq 0$ $(i = 1, 2, \ldots, m)$, it follows from the convexity assumption (8) that

$$\sum_{i=1}^{m} (x_i')^P \cdot (v_i^Q - v_i^P) > 0$$

and

$$\sum_{i=1}^{m} (x_i')^Q \cdot (v_i^P - v_i^Q) > 0.$$

Multiplying by u and λ respectively (both positive) and adding, we get (cf. (22)–(23))

$$[q_1 - \lambda \cdot (x_1')^Q] \cdot (v_1^Q - v_1^P) > 0$$

or, since $v_1^Q = 0$, $v_1^P < 0$,

$$q_1 > \lambda \cdot (x_1')^Q. \tag{24}$$

Now, by (8) we have with Q as the fixed point ($v_i^0 = v_i^Q$)

$$(x_1')^Q \cdot (v_1 - v_1^Q) + \sum_{i=2}^{m} (x_i')^Q \cdot (v_i - v_i^Q) > 0 \tag{25}$$

---

[1] In the two-factor case (Fig. 26) the solution Q follows directly from the two side conditions; there is no range of choice and hence no need for minimization.

for any point $(v_1, v_2, \ldots, v_m)$, distinct from Q. (25) combines with (23)–(24) to give

$$\sum_{i=1}^{m} q_i \cdot (v_i - v_i{}^Q) > 0$$

provided that $v_1 - v_1{}^Q \geq 0$, i.e., for any non-negative v_1. In other words, whereas c is less at P than at any other point of the isoquant, including Q, point Q is a global minimum in the region $v_1 \geq 0$ and therefore represents the optimal factor combination provided that it turns out to be feasible ($v_i{}^Q \geq 0$ also for $i = 2, 3, \ldots, m$) and efficient. This boundary solution is characterized by

$$\frac{q_1}{x_1'} > \lambda = \frac{q_i}{x_i'} \qquad (i = 2, 3, \ldots, m);$$

the interpretation is that the first factor will not be used because the partial marginal cost with respect to v_1 at the point is greater than partial marginal cost in any other factor direction so that substitution of v_1 for other inputs along the isoquant would only raise the cost of producing the given output.

Suppose, however, that $v_2{}^Q < 0$ so that Q is not feasible. Now set $v_1 = v_2 = 0$ and minimize c over again to get a new solution R, determined by $x(v_1{}^R, v_2{}^R, \ldots, v_m{}^R) = \bar{x}$, $v_1{}^R = v_2{}^R = 0$, and the conditions

$$q_i = \mu \cdot (x_i')^R \qquad (i = 3, 4, \ldots, m), \tag{26}$$

μ being a Lagrange multiplier. Then—assuming that $(x_i')^R \geq 0$ for all i—convexity implies

$$\sum_{i=1}^{m} (x_i')^R \cdot (v_i{}^P - v_i{}^R) > 0$$

$$\sum_{i=1}^{m} (x_i')^P \cdot (v_i{}^R - v_i{}^P) > 0,$$

which with (22) and (26) leads to

$$[q_1 - \mu \cdot (x_1')^R] \cdot (-v_1{}^P) > [q_2 - \mu \cdot (x_2')^R] \cdot v_2{}^P. \tag{27}$$

Similarly,

$$\sum_{i=1}^{m} (x_i')^Q \cdot (v_i{}^R - v_i{}^Q) > 0$$

and

$$\sum_{i=1}^{m} (x_i')^R \cdot (v_i{}^Q - v_i{}^R) > 0$$

combine with (23) and (26) to give

$$q_2 > \mu \cdot (x_2')^R \tag{28}$$

since $v_2{}^Q < 0$. For $v_1{}^P < 0$ and $v_2{}^P > 0$ as assumed, (27)–(28) lead to

$$q_1 > \mu \cdot (x_1')^R. \tag{29}$$

By (8) we also have

$$\sum_{i=1}^{m} (x_i')^R \cdot (v_i - v_i^R) > 0$$

from which, using (26), (28), and (29),

$$\sum_{i=1}^{m} q_i \cdot (v_i - v_i^R) > 0$$

for any point $(v_1, v_2, \ldots, v_m)$ with non-negative v_1 and v_2. Hence R represents the least-cost factor combination in the region $v_1 \geq 0$, $v_2 \geq 0$ if it is feasible ($v_i^R \geq 0$ for $i = 3, 4, \ldots, m$) and efficient. If v_3^R turns out to be negative, however, we proceed to a new solution S, setting $v_1^S = v_2^S = v_3^S = 0$, and so on.

Thus the least-cost point is a boundary solution characterized by

$$q_j > v \cdot x_j', \quad q_i = v \cdot x_i' \quad (j = 1, 2, \ldots, k; \; i = k+1, \ldots, m) \tag{30}$$

where $v_1, v_2, \ldots, v_k = 0$, and these conditions[1] are satisfied by the first of the efficient points $Q, R, S, \ldots$ which turns out to be feasible—provided that the original solution as determined by (12)–(13) has only one negative input ($v_1^P < 0$). If also $v_2^P < 0$, (30) together with the convexity assumption (8) is still sufficient for a feasible point to represent the least-cost combination, but (30) is no longer automatically satisfied by the point because (29) does not necessarily follow from (27)–(28) for $v_2^P < 0$.

The general problem of minimizing c subject to $x = \bar{x}$ and $v_i \geq 0$ can also be dealt with by Kuhn and Tucker's modified Lagrange method for solving problems of non-linear programming. The side condition $x = \bar{x}$ can be broken up into the equivalent pair of inequalities $x \geq \bar{x}$, $x \leq \bar{x}$. Then the problem can be formulated as the maximization of $-c$ subject to $x - \bar{x} \geq 0$, $\bar{x} - x \geq 0$, and $v_i \geq 0$. This is equivalent to finding a saddle point—i.e., a maximum with respect to the v_i and a minimum with respect to the Lagrange multipliers u_1 and u_2, also required to be non-negative—of the Lagrangian

$$L = -\sum_{i=1}^{m} q_i v_i + u_1 \cdot [x(v_1, v_2, \ldots, v_m) - \bar{x}] + u_2 \cdot [\bar{x} - x(v_1, v_2, \ldots, v_m)].$$

Necessary conditions for a constrained maximum of $-c$ are that there exist multipliers u_1 and u_2 such that the following necessary conditions for a saddle value of L are satisfied[2]:

---

[1] SAMUELSON (1947), pp. 69 f., states the condition in the form

$$q_i/x_i' \geq v$$

for inputs not actually used. This differs from (30) only in that it includes the special case of equality where some of the inputs which were not set $= 0$ beforehand turn out also to be zero. (For example, if Q is the least-cost point, v_2^Q may turn out to be $= 0$ so that the solutions Q and R coincide and we have not only $q_2/(x_2')^Q = \lambda$ but also $q_2/(x_2')^R = \mu$.)

[2] Cf. Appendix 1, C, conditions (15)–(16).

(a) $\dfrac{\delta L}{\delta v_i} = -q_i + (u_1 - u_2) \cdot x_i' \leq 0$ $(i = 1, 2, \ldots, m)$

(b) $v_i \geq 0$

(c) $v_i \cdot \dfrac{\delta L}{\delta v_i} = v_i \cdot [-q_i + (u_1 - u_2) \cdot x_i'] = 0$

$$(31)$$

(d) $\dfrac{\delta L}{\delta u_1} = x - \bar{x} \geq 0, \quad \dfrac{\delta L}{\delta u_2} = \bar{x} - x \geq 0$ (i.e., $x = \bar{x}$)

(e) $u_1 \geq 0, \quad u_2 \geq 0$

(f) $u_1 \cdot \dfrac{\delta L}{\delta u_1} = u_1 \cdot (x - \bar{x}) = 0, \quad u_2 \cdot \dfrac{\delta L}{\delta u_2} = u_2 \cdot (\bar{x} - x) = 0.$

Now any feasible point on the isoquant satisfies (b), (d), and (f). Conditions (a), (c), and (e) require that

$$q_i = (u_1 - u_2) \cdot x_i' \quad \text{for } v_i > 0$$
$$q_j \geq (u_1 - u_2) \cdot x_j' \quad \text{for } v_j = 0$$

for some non-negative u_1, u_2. An efficient point determined by minimizing c subject to $x = \bar{x}$ and $v_j = 0$ $(j = 1, 2, \ldots, k)$ and satisfying (30) will meet these requirements, as it is always possible to find $u_1, u_2 \geq 0$ such that their difference $u_1 - u_2$ is identified with the positive Lagrange multiplier v (cf. (30)). Such a point therefore satisfies the necessary conditions (31) [1]. It is *sufficient* for the saddle point $(v_1^0, \ldots, v_m^0, u_1^0, u_2^0)$ to represent a global maximum of $-c$ (minimum of c) that it also satisfies the condition [2]

$$L(v_1, \ldots, v_m, u_1^0, u_2^0) \leq L(v_1^0, \ldots, v_m^0, u_1^0, u_2^0) + \sum_{i=1}^{m} \left(\frac{\delta L}{\delta v_i}\right)^0 \cdot (v_i - v_i^0).$$

For $u_1 - u_2 = v > 0$ this reduces to (8) (with ≥ 0 instead of > 0), which is satisfied when the isoquant is convex.

(d) The existence of capacity limitations $v_s \leq \bar{v}_s$ is another reason why the least-cost point is not always characterized by a marginal equilibrium of the type (13). Capacities present no problem when, for technical reasons, the services of the fixed factors have to be utilized to the full; the corresponding v_s are set $= \bar{v}_s$ in the production function and hence eliminated as variables, and the capacity restrictions which are thus built into the model are automatically satisfied at any point. However, when the utilization of a fixed factor is variable (to be determined in the optimization process), v_s must be specified explicitly and the inequality $v_s \leq \bar{v}_s$ must be taken into account. If the solution to (12)–(13) fails to respect this boundary condition, it is infeasible and will have to be replaced by a boundary solution. The procedure by which a feasible least-cost point is found is very similar to that used above where the boundary conditions were $v_i \geq 0$.

---

[1] The tangency solution P (or Q) in Fig. 23 is seen to be the special case where $q_i = u \cdot x_i'$ for $i = 1, 2, \ldots, m$.

[2] Appendix 1, C, condition (17).

Fig. 27 illustrates a case of this type, v_1 representing the services of a fixed factor of capacity $\bar{v}_1$. The point of tangency[1], where $q_1/x_1' = q_2/x_2'$, is the least-cost point only if it respects the capacity limitation $v_1 \leq \bar{v}_1$; if it does not—cf. point P where $v_1^P > \bar{v}_1$ so that the solution is not feasible—the boundary condition becomes effective and we must look for another solution which minimizes cost for $v_1 = \bar{v}_1$, i.e., a solution on the boundary of the feasible region. In Fig. 27 this is obviously point Q (where c is greater than at P).

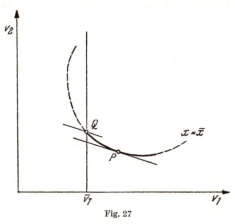

Fig. 27

More generally, let the point P represent the minimum of c on the isoquant so that

$$q_i = u \cdot (x_i')^P \qquad (i = 1, 2, \ldots, m)$$

(cf. (13)). Suppose that $v_1^P > \bar{v}_1$. To get a feasible solution set $v_1 = \bar{v}_1$ and minimize cost over again; the new solution—assumed to be non-negative—is determined by (12) and

$$q_i = \lambda \cdot (x_i')^Q \qquad (i = 2, 3, \ldots, m) \tag{32}$$

for $v_1^Q = \bar{v}_1$. Then, assuming that P and Q are both efficient points, the convexity assumption (8) again leads to

$$[q_1 - \lambda \cdot (x_1')^Q] \cdot (v_1^Q - v_1^P) > 0$$

which in turn, since $v_1^Q = \bar{v}_1 < v_1^P$, leads to

$$q_1 < \lambda \cdot (x_1')^Q, \tag{33}$$

from which we have

$$q_1 \cdot (v_1 - v_1^Q) \geq \lambda \cdot (x_1')^Q \cdot (v_1 - v_1^Q) \tag{34}$$

for all $v_1 \leq v_1^Q = \bar{v}_1$. With Q as the fixed point $(v_i^0 = v_i^Q)$, (8) combines with (32) and (34) to give

$$\sum_{i=1}^{m} q_i \cdot (v_i - v_i^Q) > 0 \quad \text{for } v_1 \leq \bar{v}_1$$

so that Q represents minimum cost in the region $v_1 \leq \bar{v}_1$, i.e., Q is the best feasible solution.

---

[1] Normally we have $q_1 = 0$ for a capacity factor so that the isocost lines are horizontal.

Q is readily seen to satisfy the Kuhn-Tucker conditions for a maximum of $-c$ subject to $x \geq \bar{x}$, $x \leq \bar{x}$, $v_1 \leq \bar{v}_1$ and $v_i \geq 0$. The necessary conditions for a saddle value of the Lagrangian

$$L = - \sum_{i=1}^{m} q_i v_i + u_1 \cdot (x - \bar{x}) + u_2 \cdot (\bar{x} - x) + u_3 \cdot (\bar{v}_1 - v_1)$$

are

(a) $-q_1 + (u_1 - u_2) \cdot x_1' - u_3 \leq 0$, $-q_i + (u_1 - u_2) \cdot x_i' \leq 0$
$$(i = 2, 3, \ldots, m)$$

(b) $v_i \geq 0$ $(i = 1, 2, \ldots, m)$

(c) $v_1 \cdot [-q_1 + (u_1 - u_2) \cdot x_1' - u_3] = 0$, $v_i \cdot [-q_i + (u_1 - u_2) \cdot x_i'] = 0$
$$(i = 2, 3, \ldots, m) \quad (35)$$

(d) $x - \bar{x} \geq 0$, $\bar{x} - x \geq 0$, $\bar{v}_1 - v_1 \geq 0$

(e) $u_1 \geq 0$, $u_2 \geq 0$, $u_3 \geq 0$

(f) $u_1 \cdot (x - \bar{x}) = 0$, $u_2 \cdot (\bar{x} - x) = 0$, $u_3 \cdot (\bar{v}_1 - v_1) = 0$.

Assuming as we did that $v_i \geq 0$ at Q, the solution Q is immediately seen to satisfy (b), (d), and (f). Clearly there exist non-negative u_1 and u_2 such that $u_1 - u_2$ can be identified with λ; this takes care of (a) and (c) for $i = 2$, $3, \ldots, m$. In order for (c) to hold for $i = 1$, (a) for $i = 1$ must be satisfied in its equational form (v_1 being $= \bar{v}_1 \neq 0$), i.e., we must have

$$u_3 = -q_1 + (u_1 - u_2) \cdot x_1'$$

which must be ≥ 0 in order to satisfy (e). It follows from (33) that this is so; hence conditions (35) are all satisfied[1].

E. The Expansion Path and the Cost Function

(a) For parametric $\bar{x}$, the locus of least-cost points determines the expansion path, along which the cost function is defined, as a one-dimensional curve in m-dimensional factor space, each v_i being given implicitly as a

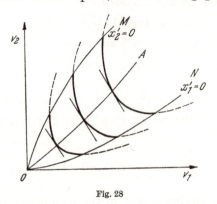

Fig. 28

---

[1] Again, the *sufficient* condition for the saddle point satisfying (35) to represent a global minimum of cost—i.e., condition (17) in Appendix 1, C—is easily shown to be equivalent to (8), the convexity assumption.

function of $\bar{x}$ by (12) and the first-order minimum conditions. When the region of substitution is contained in the non-negative factor space (i.e., $v_i \geq 0$ for all points where $x_i' \geq 0$ for $i = 1, 2, \ldots, m$), the path is determined by (12)–(13)—i.e., characterized by isocost lines being tangent to the isoquants—as long as there are no effective capacity limitations. OA in *Fig. 28* is such a path for both factor prices positive. If one of the prices is zero, the path will coincide with the boundary of the region of substitution; for $q_1 = 0$ it will be ON since (13) gives $x_1' = 0$. (Geometrically, the isocost lines become horizontal.)

The expansion path, in determining the v_i as functions of the parameter $\bar{x}$, transforms the linear cost expression (total variable cost) into a function of x,

$$c = \sum_{i=1}^{m} q_i v_i = \sum_{i=1}^{m} q_i \cdot v_i(x) = c(x).$$

The cost differential, taken at a point on the expansion path (i.e., at a point satisfying (12)–(13)), is

$$dc = \sum_{i=1}^{m} q_i \, dv_i = \sum_{i=1}^{m} (u \cdot x_i') \cdot dv_i = u \cdot dx$$

so that the multiplier u, to be interpreted as the partial marginal cost in every factor direction, is also equal to marginal cost as defined along the expansion path[1].

When the production function is homogeneous of degree one, the expansion path will be a straight line through the origin: the marginal productivities will be functions of the $m-1$ relative factor proportions v_i/v_1 $(i = 2, 3, \ldots, m)$, which together with u are uniquely determined by (13) so that they are independent of the level of output $\bar{x}$. The corresponding cost function will have the linear form $c = \bar{c} \cdot x$ where $\bar{c} \; (= u)$ is the constant marginal and average cost.

In the non-homogeneous case, nothing general can be said about the shape of the cost curve. However, if the production function in m specified variable inputs is derived from a function $x = x(v_0, v_1, \ldots, v_m)$ in which some (specifiable) factor or factor complex v_0 is held constant so that we have

---

[1] This is an illustration of the fact that Lagrange multipliers often have a significant economic interpretation (in this case, that of a "shadow price" associated with an additional unit of output, the cost amount saved by not producing it); cf. Appendix 1, A. The result could also have been derived more directly from the minimization of the Lagrangian $L = c + u \cdot (\bar{x} - x)$; the necessary condition is for any $\bar{x}$

$$dL = dc - u \cdot \sum_{i} x_i' dv_i = 0$$

where a shift in the parameter $\bar{x}$ leads to changes in the v_i satisfying the side condition, i.e.,

$$\sum_{i} x_i' dv_i = d\bar{x}$$

so that we have

$$dc - u \cdot d\bar{x} = 0.$$

$$x = x(\bar{v}_0, v_1, \ldots, v_m) = f(v_1, v_2, \ldots, v_m),[1]$$

the function f is likely to be characterized by diminishing returns to the specified variable inputs[2], especially if the production function in $m+1$ inputs is homogeneous of the first degree for all inputs (including v_0) variable. Such cases will often lead to the familiar cost pattern characterized by U-shaped marginal cost.

In the simplest possible case ($m=1$) where the production function is

$$x = x(\bar{v}_0, v_1) = f(v_1),$$

the expansion path in a (v_0, v_1) diagram is $v_0 = \bar{v}_0$ and the cost function (total variable cost) will have the shape of the inverse of the production curve $x = f(v_1)$:

$$c = q_1 v_1 = q_1 f^{-1}(x) = c(x).$$

Diminishing returns in the general sense of decreasing average productivity x/v_1 (at least from a certain point) will be reflected in increasing average cost, $c/x = q_1 v_1 / x$. More specifically, if $x = x(v_0, v_1)$ is homogeneous of degree one and $x = f(v_1)$ conforms to the law of variable proportions, the cost picture will be that of *Fig. 29*.

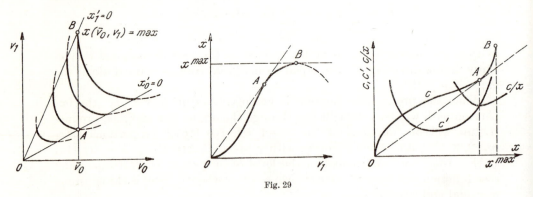

Fig. 29

Average cost c/x will have a minimum at the point A where average productivity x/v_1 has a maximum (i.e., where $x_0' = 0$), and marginal cost

$$c'(x) = \frac{d(q_1 v_1)}{dx} = \frac{q_1}{f'(v_1)}$$

will also follow a U-shaped curve, tending to infinity as x approaches its maximum (point B, the capacity limit in terms of x imposed by the fixity of v_0).

More generally, let the non-homogeneous production function

$$x = f(v_1, v_2, \ldots, v_m)$$

be derived from the homogeneous function

---

[1] v_0 may represent, for example, a capacity factor which for technical reasons has to be utilized to the full so that the production function $x = f(v_1, \ldots, v_m)$ has a "built-in" capacity limitation.

[2] Cf. Ch. IV, B above and Appendix 2.

$$x = x(v_0, v_1, \ldots, v_m)$$

for $v_0 = \bar{v}_0$. Then the expansion path no longer represents partial variation of a single input and the cost curve cannot be based on the inverse of a production curve. Instead, the expansion path is determined implicitly by

$$q_i - u \cdot f_i'(v_1, v_2, \ldots, v_m) = 0 \qquad (i = 1, 2, \ldots, m)$$

$$f(v_1, v_2, \ldots, v_m) = \bar{x} \tag{36}$$

for parametric $\bar{x}$.

Although the relative factor proportions $v_2/v_1, \ldots, v_m/v_1$ will not be constant along this path (in which case the problem would reduce to a two-factor analysis where v_1 would now represent a homogeneous factor complex), the cost picture will resemble that of Fig. 29. Any set of constant relative factor proportions v_i/v_0 $(i = 1, 2, \ldots, m)$ is represented by a half-line through the origin in $(m+1)$-dimensional factor space; each half-line is characterized by constant average cost. Assuming that the region of substitution as defined by $x_0' \geq 0$, $x_i' \geq 0$ $(i = 1, 2, \ldots, m)$ is contained in the positive $(m+1)$-dimensional factor space, the efficient half-lines will be positively inclined and one of them, as determined by minimizing cost subject to $\bar{x} = x(v_0, v_1, \ldots, v_m)$ and thus characterized by

$$x_0'(v_0, v_1, \ldots, v_m) = 0$$

$$q_i - \lambda \cdot x_i'(v_0, v_1, \ldots, v_m) = 0 \qquad (i = 1, 2, \ldots, m) \tag{37}$$

where λ is a Lagrange multiplier, represents the locus of least-cost (and least average cost) points. For $\bar{x}$ such that (37) is satisfied for $v_0 = \bar{v}_0$, i.e., at the point where the half-line intersects the plane $v_0 = \bar{v}_0$, λ can be identified with the multiplier u of (36). This means that the point lies on the expansion path (36) and is characterized by marginal cost being equal to average cost. At any other point on the plane, and thus at any other point of the expansion path (36), average cost is higher, assuming convexity as usual. If it is also assumed that $c(0) = 0$, it follows that the total cost function—and hence also average and marginal cost—will have the same general shape as in Fig. 29.

A similar cost picture can be derived from the assumption that the elasticity of output with respect to scale, ε—that is, the elasticity of x with respect to a proportional variation of the variable inputs $v_1, \ldots, v_m$—decreases from values greater than 1 through 0 to negative values as output increases along the expansion path (which, as we have seen, does not have to be linear for ε to be defined at any point of the path)[1]. This pattern, which may be said to summarize the generalized law of variable proportions, is deducible from a few simple assumptions if the production function is derived from a homogeneous function $x = x(v_0, v_1, \ldots, v_m)$ for $v_0 = \bar{v}_0$, but it may also occur in cases where the non-homogeneity of the production function $x = f(v_1, v_2, \ldots, v_m)$ cannot meaningfully be ascribed to the constancy of an unspecified input.

[1] Cf. Ch. IV, B above.

At any point on the expansion path we have, by (13),

$$c = \sum_{i=1}^{m} q_i v_i = u \cdot \sum_{i=1}^{m} v_i f_i'$$

where $u = c'$ and, by (4), $\sum_i v_i f_i' = \varepsilon \cdot x$. Thus we get[1]

$$\varepsilon = c / (x \cdot c') = \frac{dx}{dc} \cdot \frac{c}{x}, \tag{38}$$

that is, ε is not only the elasticity of production but also the flexibility of cost (the reciprocal elasticity of the cost function, or the elasticity of output with respect to cost along the expansion path). When ε is a known function of x along the expansion path, (38) is a differential equation which can be solved to give the cost function $c = c(x)$. In the present case, $\varepsilon(x)$ is incompletely specified, but the assumed properties of the function are sufficient to determine the main characteristics of the cost function. It follows from (38) that for $\varepsilon \gtreqless 1$ we have

$$c' \lesseqgtr \frac{c}{x} \quad \text{and} \quad \frac{d(c/x)}{dx} = \frac{x \cdot c' - c}{x^2} \lesseqgtr 0.$$

Thus average cost will be decreasing in the initial phase of increasing returns to scale ($\varepsilon > 1$), increasing in the subsequent phase of decreasing returns ($0 < \varepsilon < 1$), and be a (global) minimum at the point of constant returns to scale ($\varepsilon = 1$), where it is intersected from below by the marginal cost curve. As to marginal cost, we have for $0 < \varepsilon \leq 1$, if ε is assumed to decrease monotonically in this interval,

$$\frac{d\varepsilon}{dx} = \frac{x \cdot (c')^2 - c \cdot (x \cdot c'' + c')}{(x \cdot c')^2} < 0$$

and hence

$$c'' > \frac{c'}{c} \cdot \left(c' - \frac{c}{x} \right), \quad \text{where } c' - \frac{c}{x} \geq 0 \text{ for } \varepsilon \leq 1$$

so that marginal cost is increasing in this phase; by (38), $c' \to \infty$ as output expands towards the point where ε becomes zero. Finally, the additional assumption $c(0) = 0$ guarantees that there will be an initial phase of decreasing marginal cost. It is geometrically obvious that a total cost curve starting at the origin must have this property if c/x is to have a minimum for some positive x; more rigorously, expansion of the cost function by Taylor's series,

$$c(0) = c(x) - x \cdot c'(x) + \frac{x^2}{2} \cdot c''(\vartheta x)$$

for some positive fraction ϑ, shows that $c''(\vartheta x)$, and hence also $c''(x)$ for small x, will be negative, recalling that $c(0) = 0$ and that $c - x \cdot c' > 0$ for $\varepsilon > 1$. It follows that the marginal cost curve is likely to be U-shaped,

---

[1] Cf. SCHNEIDER (1934), pp. 42 f., and FRISCH (1956), pp. 163 f.

having a minimum ($c'' = 0$) for some x in the phase of increasing returns to scale ($\varepsilon > 1$).

(b) The expansion path and the shape of the cost function will be somewhat different if cost for parametric $\bar{x}$ is to be minimized subject not only to the production function but also to explicit capacity limitations in inequality form

$$v_s \leq \bar{v}_s,$$

where v_s represents the divisible services of some fixed factor, available up to the maximum amount $\bar{v}_s$ per period. As long as the least-cost point as determined by (12)–(13) does not violate the capacity restriction, there is nothing to add. However, for sufficiently large $\bar{x}$ the path will hit the plane $v_s = \bar{v}_s$ and further expansion along the path would give least-cost solutions in which $v_s > \bar{v}_s$. Hence the expansion will be deflected at the critical point (where v_s becomes $= \bar{v}_s$) and follow another path, now on the plane $v_s = \bar{v}_s$. [1]

In the two-factor case of *Fig. 30*, where v_1 is a capacity factor subject to $v_1 \leq \bar{v}_1$, parametric minimization of cost obviously leads to the kinked expansion path OQB (or OAB if no positive price is associated with v_1), whereas it would have been PB—i.e., the line $v_1 = \bar{v}_1$—if the capacity factor v_1 had not been divisible.

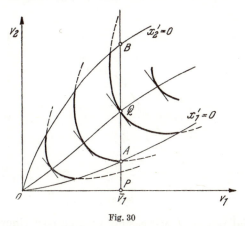

Fig. 30

Up to the point of deflexion, the resulting cost function will be lower than total (variable) cost as defined along $v_1 = \bar{v}_1$; divisibility makes it possible to produce any amount of output within this interval more cheaply than along PQ, where the marginal productivity of v_1 is "too small" (between P and A, even negative). Assuming that the cost function along $v_1 = \bar{v}_1$ resembles that corresponding to the law of variable proportions, total variable cost as defined along OQB will have the shape illustrated by *Fig. 31*. The effect of divisibility is to smooth out the initial phase of increasing returns. The cost function has no kink at the point where the expansion path is deflected since, as we have seen, marginal cost at the point is equal to the partial marginal cost in the direction of v_2.

---

[1] Cf. Ch. IV, D above.

If the production function is homogeneous of degree one, $0Q$ (or $0A$) becomes a straight line and the corresponding segment of the cost function will be linear. This case is shown in *Fig. 32*, assuming that $q_1 = 0$ and that

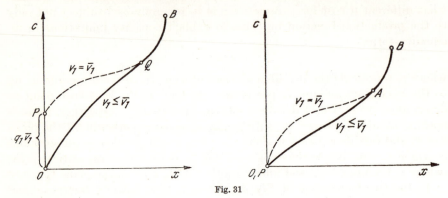

Fig. 31

the law of variable proportions holds for partial variation of v_2. For $v_1 = \bar{v}_1$ the cost picture would be similar to that shown in Fig. 29; for $v_1 \leq \bar{v}_1$ we have the kinked expansion path $0AB$ where the point of deflexion coincides with the point of minimum average cost in the case of indivisibility[1].

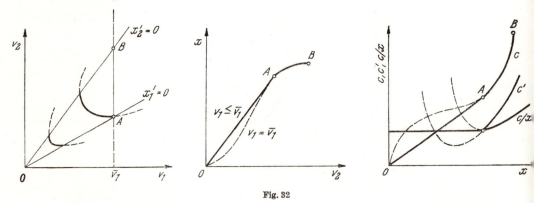

Fig. 32

Thus the initial phase of decreasing average cost (increasing returns to scale, $\varepsilon > 1$) is replaced by one of constant average and marginal cost, $0A$ (where $\varepsilon = 1$). Beyond A, total cost is upward bending because the optimal factor combination cannot be maintained when the capacity factor is fully utilized.

On suitable assumptions, this result holds for any number of inputs. Minimizing

$$c = \sum_{i=2}^{m} q_i v_i$$

---

[1] For $v_1 = \bar{v}_1$, $c/x = q_2 v_2 / x$ is a minimum for $x_1' = 0$, i.e., $x_2' = x/v_2$; cf. above.

Note that marginal cost is kinked (but continuous) at A, c'' being $= 0$ along the linear segment but positive for movements beyond A; average cost is not kinked since the linear segment is tangent to c/x at the latter's minimum.

(assuming $q_1 = 0$) subject to

$$x(v_1, v_2, \ldots, v_m) = \bar{x}$$

leads to the conditions

$$x_1'(v_1, v_2, \ldots, v_m) = 0$$

$$q_i - u \cdot x_i'(v_1, v_2, \ldots, v_m) = 0 \qquad (i = 2, 3, \ldots, m)$$

which, assuming homogeneity, determine the $(m-1)$ relative factor proportions—independent of $\bar{x}$—and the constant marginal cost, u. At the critical level where v_1 becomes equal to $\bar{v}_1$ the expansion path is deflected, now to be determined by minimizing c subject to $x = \bar{x}$ with $v_1 = \bar{v}_1$. This gives

$$q_i - \lambda \cdot x_i'(\bar{v}_1, v_2, \ldots, v_m) = 0 \qquad (i = 2, 3, \ldots, m)$$

which, with the production function, determine $v_2, v_3, \ldots, v_m$ and marginal cost λ (no longer constant) as functions of $\bar{x}$. At the point of deflexion we have $\lambda = u$ (both equal to q_i/x_i', $i = 2, 3, \ldots, m$, where $v_1 = \bar{v}_1$), which in turn represents minimum average cost[1]. Hence, if the cost function for $v_1 = \bar{v}_1$ conforms to the pattern of the law of variable proportions (Fig. 29), we get the pattern of Fig. 32 for $v_1 \leq \bar{v}_1$.

F. The Optimum Level of Output

(a) For a given cost function, the optimum level of output is now determined by maximizing (gross) profit

$$z = p \cdot x - c(x)$$

(where fixed cost elements can be disregarded since a constant term does not affect the solution). The *necessary* condition for maximum is

$$\frac{dz}{dx} = p - \frac{dc(x)}{dx} = 0 \text{ or } p = c', \tag{39}$$

i.e., price (more generally: marginal revenue) must equal marginal cost,

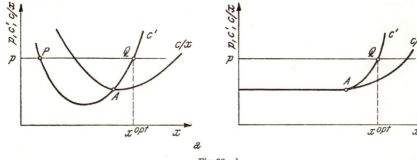

Fig. 33 a–b

as illustrated in *Fig. 33a–b* for the two different types of cost functions corresponding to Fig. 29 and 32 respectively.

---

[1] Cf. the explanation of (36)–(37) above.

A *sufficient* condition for a point satisfying (39) to represent a (local) maximum is that

$$\frac{d^2z}{dx^2} = -\frac{d^2c\,(x)}{dx^2} < 0 \text{ or } c'' > 0\,, \tag{40}$$

i.e., marginal cost must be increasing at the point[1]. This criterion makes it possible to distinguish between those solutions to (39) which represent maxima (such as Q in Fig. 33 a–b) and those which do not (point P).

A solution to (39) must be required to give a positive (gross) profit or it would be better to produce nothing at all[2]; that is, it must also satisfy

$$z = p \cdot x - c\,(x) > 0\,.$$

With (39) this condition gives

$$p = c' > \frac{c}{x}\,;$$

this means that we must look for the optimum point in the region where marginal cost is higher than average cost (and the latter therefore rising), i.e.—by (38)—in the region of decreasing returns to scale, $\varepsilon < 1$. [3] With the type of cost function shown in Fig. 33 a, where a U-shaped marginal cost curve intersects average cost at the latter's minimum, this means the region to the right of the point of intersection, A. (It is possible to have (39)–(40) satisfied at a point to the left of A, but only if p is lower than minimum average cost, in which case gross profit is clearly negative.) If ε decreases monotonically in this region, c' will rise monotonically, tending to infinity as $\varepsilon \to 0$;[4] it follows that a point in the region satisfying (39) will automatically satisfy the sufficient condition (40) and, moreover, that the solution is unique and therefore represents a global maximum of profit[5] [6].

---

[1] If the selling price p is not constant but decreases with x, this result does not hold. The modification of (39)–(40) necessary to suit this case is obvious.

[2] Net profit, being gross profit minus a constant (fixed costs), will have its maximum at the same point as gross profit. In the short run, profit maximization does not imply the requirement that net profit is to be positive: a point where gross profit z is positive is clearly better than producing nothing, in which case there is no revenue to cover part of the fixed cost.

[3] Cf. FRISCH (1956), pp. 171 ff.

[4] Cf. Ch. IV, E above.

[5] More rigorously, expanding the total cost function by Taylor's series from the optimum point x_0 (where $p = c'(x_0)$) we have

$$c\,(x) = c\,(x_0) + (x - x_0) \cdot c'\,(x_0) + \frac{(x - x_0)^2}{2} \cdot c''$$

where c'' is taken at some intermediate point. We know that $c'' > 0$ for all x und x^0 in the region $\varepsilon < 1$. It follows that

$$c\,(x) - c\,(x_0) > (x - x_0) \cdot c'\,(x_0)$$

so that

$$z\,(x_0) - z\,(x) = p \cdot x_0 - p \cdot x - c\,(x_0) + c\,(x) > -p\,(x - x_0) + (x - x_0) \cdot c'\,(x_0) = 0\,,$$

i.e.,

$$z\,(x_0) > z\,(x) \text{ for all } x \neq x_0 \text{ in the region.}$$

[6] Other types of cost functions may display multiple solutions to (39) satisfying (40), in which case the optimal solution is determined as the *maximum maximorum*.

The same conclusions apply to the cost function of Fig. 33 b, which differs from Fig. 33 a only in that $\varepsilon = 1$ all the way up to point A (reflecting the proportionate expansion of all inputs of a homogeneous production function up to the point where the linear expansion path is deflected by an explicit capacity limitation); beyond this point the marginal cost curve has the shape of Fig. 33 a.

It follows that, if the marginal cost curves of Fig. 33 a–b are derived from the same homogeneous production function for $v_1 = \bar{v}_1$ and $v_1 \leq \bar{v}_1$ respectively, it makes no difference to the determination of the optimum level of output whether the capacity limitation is stated in equality or inequality form: the most profitable point (Q) will in either case be to the right of A, that is, a point at which the capacity of the fixed factor is fully utilized $(v_1 = \bar{v}_1)$.

In the absence of a capacity limit, homogeneity would lead to constant marginal and average cost for any value of x and output would be indeterminate[1]; (39) would not determine x and (40) could never be satisfied. However, it is not easy to see how this could possibly happen: there must always be some factor—specified or unspecified—which limits the amount of output that can be produced per period within a given plant.

(b) Instead of first determining the cost function and then maximizing revenue minus cost, the profit expression

$$z = p \cdot x(v_1, v_2, \ldots, v_m) - \sum_{i=1}^{m} q_i v_i$$

may be maximized directly. With all v_i as independent variables[2] the *necessary* condition is

$$dz = \sum_{i=1}^{m} \frac{\delta z}{\delta v_i} dv_i = 0$$

or

$$\frac{\delta z}{\delta v_i} = p \cdot x_i' - q_i = 0 \qquad (i = 1, 2, \ldots, m) \tag{41}$$

which, with the production function, determines x and the v_i. Since $u = q_i/x_i' = c'$, conditions (41) are equivalent to the necessary conditions for minimum cost and for maximum profit on the expansion path, (13) and (39), and therefore lead to the same optimum point. Another interpretation of (41) is that, at the equilibrium point, the price of a marginal unit of each input is equal to the value of its marginal contribution to output.

---

[1] Unless the selling price is a decreasing function of x so that marginal revenue is falling.

[2] It is assumed that the capacity limitation is built into the production function, being represented by a parameter $\bar{v}_0$ in the production function (to be interpreted as the indivisible services of a fixed factor). If the fixed factor's services are divisible, they must be specified in the catalogue of input variables—i.e., represented by one of the v_i in $x = x(v_1, \ldots, v_m)$, say, v_1—and a capacity constraint must be added $(v_1 \leq \bar{v}_1)$. For reasons stated above we may set $v_1 = \bar{v}_1$ from the outset and maximize z with respect to the $(m-1)$ variables $v_2, \ldots, v_m$. The following argument is not materially affected by this adjustment.

A *sufficient* condition for a point satisfying (41) to represent a maximum is that we have at the point

$$d^2z = d\left[\sum_{i=1}^{m}(p\,x_i' - q_i)\,dv_i\right] = p \cdot \sum_{i=1}^{m}\sum_{j=1}^{m} x_{ij}''\,dv_i\,dv_j < 0$$

or

$$\sum_{i=1}^{m}\sum_{j=1}^{m} x_{ij}''\,dv_i\,dv_j < 0 \qquad (42)$$

for *all* variations dv_i, dv_j from the point[1]. Again, condition (42) can be shown to contain the sufficient conditions for least cost and maximum profit in the two-step analysis, (14) and (40). Clearly, when (42) is satisfied for all variations (all v_i being independent variables), $\sum_i\sum_j x_{ij}''\,dv_i\,dv_j$ is also negative subject to $\sum_i x_i'\,dv_i = 0$, i.e., for movements along the isoquant, so that (14) is satisfied. Moreover, at a point on the expansion path we have

$$c' \cdot x_i' = q_i \qquad (i = 1, 2, \ldots, m)$$

and variations from the point along the path will have to satisfy the equations $d(c'\,x_i') = dq_i$ or

$$c' \cdot \sum_{j=1}^{m} x_{ij}''\,dv_j + x_i' \cdot c'' \cdot dx = 0 \qquad (i = 1, 2, \ldots, m)$$

which, by multiplication by dv_i and summation over i—recalling that $\sum_i x_i'\,dv_i = dx$—leads to

$$c'' = -\frac{c'}{dx^2} \cdot \sum_{i=1}^{m}\sum_{j=1}^{m} x_{ij}''\,dv_i\,dv_j,$$

i.e., $c'' > 0$ (condition (40)) when (42) is satisfied[2].

---

[1] The quadratic form (42) is negative definite if and only if

$$(-1)^k \cdot \begin{vmatrix} x_{11}''x_{12}'' & \ldots\ldots & x_{1k}'' \\ x_{21}'' & & \cdot \\ \cdot & & \cdot \\ \cdot & & \cdot \\ \cdot & & \cdot \\ x_{k1}'' & \ldots\ldots\ldots & x_{kk}'' \end{vmatrix} > 0 \qquad \text{for } k = 1, 2, \ldots, m$$

which is therefore an alternative formulation of the restriction (42) which the production function must satisfy in order for the point to represent a maximum. The second-order condition is given in this form by many authors, e.g. FRISCH (1956), p. 177.

[2] With only one variable factor, (42) becomes

$$\frac{d^2x(v)}{dv^2} = x'' < 0,$$

i.e., the maximum point must be somewhere on the declining branch of the marginal productivity curve. This, of course, is equivalent to the condition

(c) The second-order maximum condition makes it possible to determine the direction of the change in x brought about by a marginal change in the price parameter, p. Differentiating the equilibrium condition (39) with respect to p we have

$$1 - \frac{d^2c\,(x)}{dx^2} \cdot \frac{dx}{dp} = 0 \quad \text{or} \quad \frac{dx}{dp} = \frac{1}{c''}$$

which must hold if the optimal value of x is continuously adjusted to changes in p. Combined with (40), this gives $dx/dp > 0$ (as was geometrically obvious from Fig. 33).

The same result can be derived from the equivalent conditions (41)–(42). Differentiating (41) we have, in differential form,

$$p \cdot \sum_{j=1}^{m} x_{ij}'' \, dv_j + x_i' \cdot dp = 0 \qquad (i = 1, 2, \ldots, m);$$

solving for the dv_j in terms of dp and substituting in $dx = \sum_j x_j' dv_j$—the production function being also part of the equilibrium conditions—we get dx in terms of dp. The same result, however, can be reached by a short cut. Multiplication by dv_i and summation over i gives

$$p \cdot \sum_{i=1}^{m} \sum_{j=1}^{m} x_{ij}'' \, dv_i \, dv_j + \left(\sum_{i=1}^{m} x_i' dv_i \right) \cdot dp = 0$$

where the dv_i must satisfy $\sum_i x_i' \, dv_i = dx$. By (42), the first term is negative so that we have $dx \cdot dp > 0$, that is, a rise in p leads to greater output.

Similarly, output will rise if the marginal cost curve is lowered because of factor price changes. As we have seen, a *proportional* change in all factor prices will not affect the optimal factor combination for any given x (cf. (12)–(13)) but will change marginal cost $c' = u = q_i/x_i'$ in the same proportion. Write $q_i = k_i \cdot q$ $(i = 1, 2, \ldots, m)$ where the k_i are constants. Then (39) can be written in the form

$$p - c'\,(x, q) = 0;$$

differentiating with respect to the factor price parameter q we have

$$- \left(c'' \cdot \frac{dx}{dq} + \frac{\delta c'}{\delta q} \right) = 0$$

where $\delta c'/\delta q = k_i/x_i' > 0$ for any i so that we get

$$\frac{dx}{dq} = - \frac{\delta c'/\delta q}{c''} = \frac{-k_i}{c'' \cdot x_i'}$$

which is negative by (40).

$$\frac{d^2c}{dx^2} = c'' > 0$$

since marginal cost $c' = q/x'$ so that we have

$$c'' = \frac{-q}{(x')^3} \cdot x''.$$

Similarly, for *partial* variation in one of the factor prices, e. g. q_m, we have

$$\frac{dx}{dq_m} = -\frac{\delta c' / \delta q_m}{c''}$$

which is negative if $\delta c' / \delta q_m = \delta c' (x, q_m) / \delta q_m > 0$ at the equilibrium point. Although this cannot be generally established[1], it seems reasonable to assume that a decrease in one of the factor prices will lower marginal cost for any given level of output. A fall in q_m will reduce total cost if the factor combination is kept unaltered; adjusting the factor combination to the new optimum position (x being kept constant $= \bar{x}$) will reduce cost still further. Thus, for any positive $\bar{x}$ total cost will be lowered; this means that, if $c(0) = 0$, there will be a tendency toward lower marginal cost.

(d) As we have seen, the point of maximum profit is normally characterized by the fixed factor being fully utilized, not only when its services are indivisible but also when the capacity restriction has the inequality form $v_1 \leq \bar{v}_1$. Profit is at a maximum somewhere on the rising branch of the marginal cost curve and it is the capacity limitation which eventually causes the rise in marginal cost, thus representing an effective limit to the expansion of output and profit.

An increase in the plant's capacity would shift the relevant part of the marginal cost curve, and thus also the optimum point, to the right[2]. The resulting increase in profit represents the highest price the firm would be willing to pay for the additional capacity. This is the basis of shadow pricing: the shadow price, y, associated with the fixed factor is defined as the marginal profit that could be obtained from an additional unit of the fixed factor's services, were it available (or, which comes to the same thing, as the profit forgone by not using the marginal unit of available capacity)[3].

The displacement of equilibrium due to a marginal change in capacity can be traced by varying the parameter $\bar{v}_1$ in the optimization model. For given capacity $\bar{v}_1$—fully utilized so that $v_1 = \bar{v}_1$—the input variables are determined by the necessary equilibrium conditions

$$p \cdot x_i' (\bar{v}_1, v_2, \ldots, v_m) - q_i = 0 \qquad (i = 2, 3, \ldots, m); \qquad (43)$$

the resulting output follows from the production function

$$x = x (\bar{v}_1, v_2, \ldots, v_m) \tag{44}$$

and (gross) profit becomes

$$z = px - q_1 \bar{v}_1 - \sum_{i=2}^{m} q_i v_i \tag{45}$$

(where q_1 will normally be zero)[4]. Differentiating (43) with respect to $\bar{v}_1$ we have

---

[1] Cf. Ch. IV, D above.

[2] For example, in Fig. 32, the phase of increasing average cost (AB) will start at a larger value of x as point A moves further away from the origin.

[3] Cf. Ch. III, A and B.

[4] $q_1 > 0$ means that a positive cost is incurred by *using* (as distinct from buying) a unit of v_1; $q_1 \bar{v}_1$ must not be confused with the historically determined fixed cost associated with the firm's fixed capital equipment.

$$p \cdot x_{i1}{}'' + p \cdot \sum_{j=2}^{m} x_{ij}{}'' \cdot \frac{dv_j}{d\bar{v}_1} = 0 \qquad (i = 2, 3, \ldots, m) \qquad (46)$$

which may be solved for the $dv_j/d\bar{v}_1$ to give the displacement of the equilibrium values of $v_2, \ldots, v_m$ due to a marginal change in $\bar{v}_1$, assuming that the inputs are adjusted optimally to the changing capacity. The displacement of x and z follows by substitution in

$$\frac{dx}{d\bar{v}_1} = x_1{}' + \sum_{j=2}^{m} x_j{}' \cdot \frac{dv_j}{d\bar{v}_1} \qquad (47)$$

and

$$\frac{dz}{d\bar{v}_1} = p \cdot \frac{dx}{d\bar{v}_1} - q_1 - \sum_{j=2}^{m} q_j \cdot \frac{dv_j}{d\bar{v}_1} \qquad (48)$$

which result from differentiation of (44) and (45). However, if we are interested only in $dz/d\bar{v}_1$, there is no need to solve (46): equations (47)–(48) give

$$\frac{dz}{d\bar{v}_1} = px_1{}' - q_1 + \sum_{j=2}^{m} (px_j{}' - q_j) \cdot \frac{dv_j}{d\bar{v}_1} = px_1{}' - q_1$$

where, by (43), the coefficients of the $dv_j/d\bar{v}_1$ vanish so that the shadow price becomes

$$y = \frac{dz}{d\bar{v}_1} = px_1{}' - q_1.$$

In other words, an extra unit of $\bar{v}_1$ raises output by $x_1{}'$ units, from the value of which we must subtract the cost of using it (if any)[1]; or, alternatively,

$$y = \frac{dz}{dx} \cdot \frac{\delta x}{\delta \bar{v}_1} = (p - q_1/x_1{}') \cdot x_1{}'$$

where $p - q_1/x_1{}'$ is selling price minus partial marginal cost in the direction of v_1. [2]

---

[1] This result needs careful interpretation. It does not follow from what has been said that the current inputs $v_2, \ldots, v_m$ are kept constant when the optimum position is shifted—on the contrary, they are changed according to (46). What follows is that the changes in $v_2, \ldots, v_m$ and the change in x resulting from them cancel out as far as profit is concerned so that the marginal net effect on z is the same as if only $\bar{v}_1$ had increased. This is not surprising when it is recalled that marginal cost at an optimal point is the same in every factor direction.

[2] The shadow price can be identified with a Lagrange multiplier (cf. Appendix 1, A). Maximizing z subject to the side condition $v_1 = \bar{v}_1$, using the Lagrangian

$$L = p \cdot x (v_1, v_2, \ldots, v_m) - q_1 v_1 - \sum_{i=2}^{m} q_i v_i + u \cdot (\bar{v}_1 - v_1)$$

where u is a Lagrange multiplier, necessary maximum conditions are

$$\frac{\delta L}{\delta v_1} = px_1{}' - q_1 - u = 0, \qquad \frac{\delta L}{\delta v_i} = px_i{}' - q_i = 0 \qquad (i = 2, 3, \ldots, m)$$

$$\left(\text{or, alternatively, } dL = \sum_{i=1}^{m} (px_i{}' - q_i) dv_i - u dv_1 = dz - u dv_1 = 0 \right)$$

When the capacity is fully utilized at the optimum point, the shadow price is positive since otherwise profit could be increased by leaving part of $\bar{v}_1$ idle (equivalent to a decrease in capacity), contradicting the assumption that profit is at a maximum for $v_1 = \bar{v}_1$[1]. On the other hand, had v_1 been less than $\bar{v}_1$ at the optimum point, y would be zero since a marginal change in capacity would leave the equilibrium position and thus also profit unaltered; this reflects the intuitively obvious fact that the utilization of a fixed factor is a free good up to capacity.

When the production function is homogeneous of degree one, we have by Euler's theorem

$$x = v_1 x_1' + \sum_{i=2}^{m} v_i x_i'$$

identically. Combining this with (43) and with $v_1 = \bar{v}_1$, we have at the point of maximum profit

$$\bar{v}_1 \cdot p x_1' = px - \sum_{i=2}^{m} q_i v_i = z$$

or, if $q_1 > 0$ so that the term $q_1 \bar{v}_1$ must be included in total variable cost,

$$\bar{v}_1 \cdot (p x_1' - q_1) = px - q_1 \bar{v}_1 - \sum_{i=2}^{m} q_i v_i = z .$$

It follows not only that the shadow price—the marginal opportunity value per unit of v_1—is positive (because $z > 0$) but also that the total imputed value of the fixed factor's services, as evaluated on the basis of the shadow price, equals the total gross profit earned at the point of maximum profit[2]. Consequently, if the firm determines the scale of output on a "full-cost" basis in the sense that the selling price is required to cover average variable cost plus the unit cost associated with the fixed factor[3], this costing procedure leads to maximum profit only if the latter cost element is estimated on the basis of the shadow price. Any other cost price per unit of v_1 will lead to a less profitable scale of output[4].

---

from which we get

$$\frac{dz}{dv_1} = u = p x_1' - q_1 = y .$$

[1] This argument does not hold in the case of indivisibility where v_1 is always $= \bar{v}_1$ because less than full utilization of capacity is impossible by assumption.

[2] Cf. the concept of quasi-rent.

[3] "Full-cost" pricing in the usual sense consists in fixing the selling price on the basis of total unit cost (including some profit margin) at a predetermined level of output, whereas the problem here is to find that x for which total unit cost is exactly equal to the given price of the product.

[4] This result does not hold when the production function is not homogeneous of degree one in all m input variables.

Chapter V

More Complex Models

A. Substitution with Shadow Factors

1. "Product Shadows"

(a) The models of the previous chapters represent special cases in the sense that all inputs are in the same position: they are all substitutable for one another—continuously or otherwise—or perfectly complementary (limitational).

An obvious generalization is a "mixed" type of production model where some but not all of the inputs are mutual substitutes. To take a simple example, the technology of a process may be represented by the model

$$x = f(v_1, v_2)$$
$$v_3 = a_3 \cdot x \qquad (1)$$

or

$$x = f(v_1, v_2) = \frac{1}{a_3} \cdot v_3$$

where v_1 and v_2 are continuously substitutable whereas v_3 is a limitational input with the constant coefficient of production a_3.[1] (More generally, v_3 may be some increasing function of x.) It is an example of what has been termed a *"shadow factor"*—more specifically, a "product shadow"[2]: v_3 follows x like its shadow, depending only on the level of output regardless of what combination of v_1 and v_2 is used.

Since models of this type are characterized by the number of independent variables (degrees of freedom) being less than the number of inputs, partial variation of one input for all others constant is not possible so that marginal productivities cannot be defined in the usual way. For example, for v_2 and v_3 constant in (1), x and hence also v_1 is determined; for given values of v_1 and v_2, on the other hand, v_3 cannot vary because x is determined by the

---

[1] This case was first noticed by PARETO (1897), § 714, p. 83, and § 717, p. 85. SCHNEIDER (1934), p. 3, has suggested that models of this type have a wide range of application, actual production processes being characterized by separate groups of substitutional and limitational inputs. See also SCHULTZ (1929), p. 517, GEORGESCU-ROEGEN (1935), STIGLER (1946), pp. 364 ff., and FRISCH (1953), Ch. 12—13.

[2] FRISCH (1953), pp. 22 ff. See also SCHNEIDER (1934), p. 6.

first equation and v_3 by the second. However, it is possible to analyse the model in terms of the marginal productivities of the substitutional factors alone if they are defined as the partial derivatives of the first equation of the model:

$$f_1' = \frac{\delta f}{\delta v_1}, \qquad f_2' = \frac{\delta f}{\delta v_2},$$

it being understood that all other substitutional factors—in this case, v_2 and v_1 respectively—are kept constant whereas the limitational input v_3 is adjusted in accordance with the second equation[1].

Like other multi-equation models, (1) can be rewritten as a single production function in terms of input amounts available:

$$x = \text{Min} \; [\psi(w_1, w_2), w_3/a_3] = \varphi(w_1, w_2, w_3) \qquad (1\,\text{a})$$

where

$$\psi(w_1, w_2) = \text{Max} \, f(v_1, v_2) \quad \text{for} \quad v_1 \leq w_1, \quad v_2 \leq w_2.$$

In this formulation, where the w_i are independent variables so that partial variation is possible, the marginal productivities can be defined in the usual manner—i.e., $\varphi_i' = \delta\varphi/\delta w_i$, $i = 1, 2, 3$—only they are discontinuous at points where there is "just enough" of w_3 ($w_3 = a_3 \cdot \psi(w_1, w_2)$), i.e., at points which also belong to (1):

$$\varphi_1' = 0, \quad \varphi_2' = 0, \quad \varphi_3' = 1/a_3 \quad \text{for} \quad w_3/a_3 < \psi(w_1, w_2)$$

$$\varphi_1' = f_1', \quad \varphi_2' = f_2', \quad \varphi_3' = 0 \qquad \text{for} \quad w_3/a_3 > \psi(w_1, w_2).\,[2]$$

The optimal factor combination for given level of output is determined by minimizing variable cost subject to (1) for given $x = \bar{x}$. Since the shadow factor depends only on x, the cost of v_3 is represented by a constant term in the cost expression and the problem reduces to that of minimizing the cost of v_1 and v_2,

$$c - q_3 a_3 \bar{x} = q_1 v_1 + q_2 v_2$$

subject to

---

[1] Cf. SCHNEIDER (1934), p. 5, and FRISCH (1953), p. 23. Taking the differentials of (1) we have

$$dx = f_1' \, dv_1 + f_2' \, dv_2, \qquad dv_3 = a_3 dx$$

where the marginal productivity of v_1 is now defined as dx/dv_1 for $dv_2 = 0$ and $dv_3 = a_3 dx$, i.e., as f_1'; similarly for $dv_1 = 0$ we have $dx/dv_2 = f_2'$. The isoquant for given x is defined by $dx = dv_3 = 0$, i.e., by

$$f_1' \, dv_1 + f_2' \, dv_2 = 0$$

along which the marginal rate of substitution is

$$-dv_2/dv_1 = f_1'/f_2' > 0$$

(in the region where f_1' and f_2' are positive).

[2] GEORGESCU-ROEGEN (1935), pp. 41—43, appears to have a model of the type (1a) in mind when he defines a limitational factor by the existence of points at which all other factors have marginal productivities equal to zero (i.e., output cannot be increased without increasing the factor in question). From this he deduces a model of the type (1). However, as pointed out by KALDOR (1937), such points may also exist in the case where all inputs are substitutable (in the two-factor case, all points on the boundary of the region of substitution when this region does not cover the whole of the positive factor plane). For an input to be limitational at a point we must further require that output cannot be raised by increasing the input alone, cf. FRISCH (1953), p. 3—a condition satisfied by model (1) or (1a).

$$\bar{x} = f(v_1, v_2),$$

i.e., to a problem in the substitutional factors alone. The first-order condition for a minimum, as found by minimizing the Lagrangian expression

$$L = q_1 v_1 + q_2 v_2 + u \cdot [\bar{x} - f(v_1, v_2)],$$

is

$$\frac{q_1}{f_1'} = \frac{q_2}{f_2'} \ (= u)$$

which is the familiar equality between the ratio of factor prices and the marginal rate of substitution (i.e., the ratio of the marginal productivities as defined above). This condition together with the side condition $\bar{x} = f(v_1, v_2)$ determines the least-cost combination of v_1 and v_2. Unlike the case where all three inputs are substitutional, the position of the optimum will not be affected by a change in q_3; it depends on the prices of v_1 and v_2 only.

For parametric $x = \bar{x}$ we can in this way determine the expansion path as the locus of least-cost points, along which the cost function $c = c(x)$ is defined. The optimal level of output is, as usual, characterized by the product price being equal to marginal cost,

$$p = c'(x),$$

where the latter is

$$c'(x) = \frac{dc}{dx} = \frac{q_1 \, dv_1 + q_2 \, dv_2 + q_3 a_3 \, dx}{dx}$$

$$= \frac{u \cdot (f_1' \, dv_1 + f_2' \, dv_2)}{dx} + q_3 a_3 = u + q_3 a_3 \, ;$$

u is the marginal cost of the substitutional inputs, to which must be added a constant term representing the cost of the shadow factor per additional unit of output.[1]

The same result could have been obtained directly without introducing the cost function. The profit expression being

$$z = px - q_1 v_1 - q_2 v_2 - q_3 v_3 = (p - q_3 a_3) \cdot f(v_1, v_2) - q_1 v_1 - q_2 v_2,$$

the necessary conditions for profit to be a maximum are

$$(p - q_3 a_3) \cdot f_1' - q_1 = 0$$
$$(p - q_3 a_3) \cdot f_2' - q_2 = 0 \ [2]$$

which are seen to be equivalent to the above.

These results may have to be modified if one of the inputs represents the (divisible) services of fixed equipment. Let v_1 be such a capacity factor, constrained by the capacity limit $v_1 \leq \bar{v}_1$. As long as the solution determined

---

[1] SCHNEIDER (1934), p. 48.

[2] In SCHNEIDER's terminology, these conditions establish for each substitutional factor an equality between the factor's price and the "value of its net marginal product." The introduction of a shadow factor causes no other change in the familiar marginal equilibrium conditions than the reduction of the product price by the marginal cost of the shadow factor. See SCHNEIDER (1958), p. 212.

by the above conditions together with (1) respects the capacity restriction—whether it does or not will depend on the prices—there is nothing to add, except that q_1 is likely to be zero so that the first of the equilibrium conditions becomes $f_1' = 0$: the services of the fixed equipment, being "free," will be used up to the point of zero marginal productivity. However, if v_1 turns out to be greater than $\bar{v}_1$ so that the solution is not feasible, or if no solution exists—as will be the case when the function f is homogeneous of the first degree so that marginal cost is constant—the problem will have to be reformulated with $v_1 = \bar{v}_1$ as an additional constraint. Maximizing z with $v_1 = \bar{v}_1$ leads to the necessary condition

$$(p - q_3 a_3) \cdot f_2' - q_2 = 0$$

and the optimal point is now determined by this condition together with the production model (1) and the equation $v_1 = \bar{v}_1$, the latter having replaced the condition $\delta z / \delta v_1 = 0$.

(b) Another "mixed case" is that of a linear production model where some of the inputs are discontinuously substitutable while others are limitational shadow factors. Suppose, for example, that the product can be made in three linear activities using the same two inputs in different proportions,

$$
\begin{aligned}
x &= \lambda_1 + \lambda_2 + \lambda_3 \\
v_1 &= a_{11}\lambda_1 + a_{12}\lambda_2 + a_{13}\lambda_3 \\
v_2 &= a_{21}\lambda_1 + a_{22}\lambda_2 + a_{23}\lambda_3
\end{aligned}
$$

where λ_1, λ_2, and λ_3 are the activity levels and the a_{ij} (some of which may be zero) are the technical coefficients, and that a third input is required in the same proportion to output in each activity:

$$v_3 = a_3 \cdot x = a_3 \lambda_1 + a_3 \lambda_2 + a_3 \lambda_3 .$$

Obviously this case can be treated as an extended linear model with three inputs instead of two, the matrix of coefficients being

x	1	1	1
v_1	a_{11}	a_{12}	a_{13}
v_2	a_{21}	a_{22}	a_{23}
v_3	a_{31}	a_{32}	a_{33}

where the coefficients in the last row (those of v_3) are all equal ($a_{31} = a_{32} = a_{33} = a_3$), and the corresponding problems of optimal allocation can be solved by linear programming methods.

2. "Factor Shadows"

(a) Instead of being a limitational input depending on output, a shadow factor may be a function of one or more of the substitutional inputs—in FRISCH's terminology, a "factor shadow."[1] The simplest case is the model

$$
\begin{aligned}
x &= f(v_1, v_2) \\
v_3 &= k \cdot v_1
\end{aligned}
\tag{2}
$$

---

[1] Cf. FRISCH (1953), pp. 22 ff.

where, for some technological reason, the inputs v_3 and v_1 are required in the fixed proportion k, v_3 being a "shadow" of v_1. [1]

More generally, v_3 may be some function of v_1 and v_2:

$$x = f(v_1, v_2)$$
$$v_3 = g(v_1, v_2).$$

(2a)

Model (1) may be written in this form, the second equation being

$$v_3 = a_3 \cdot x = a_3 \cdot f(v_1, v_2).$$

The converse is not true except in the special case where the function g can be expressed as a function of f:

$$g = F(f) = F(x)$$

so that v_3 is a product shadow.

Optimization problems with production relations of type (2) can obviously be handled in terms of the substitutional input variables alone, using the second equation to eliminate the shadow factor in the cost and profit expressions. Total cost is

$$c = q_1 v_1 + q_2 v_2 + q_3 v_3 = (q_1 + k q_3) v_1 + q_2 v_2,$$

i.e., v_1 and v_3 can be treated together as a complex factor measured in terms of, say, v_1, in which case the corresponding factor price will be $q_1 + k q_3$. (If v_1 represents the services of fixed equipment, this means that the price per unit of v_1 is positive even though $q_1 = 0$: for example, the cost per machine hour is equal to the hourly wage rate when the machine has to be operated by one worker, $k = 1$.) The conditions for maximum profit are

$$p \cdot f_1' - (q_1 + k q_3) = 0$$
$$p \cdot f_2' - q_2 \qquad = 0.$$

In the case (2a) this generalizes to

$$p \cdot f_1' - \left(q_1 + \frac{\delta g}{\delta v_1} q_3\right) = 0$$

$$p \cdot f_2' - \left(q_2 + \frac{\delta g}{\delta v_2} q_3\right) = 0,$$

as we find by maximizing profit

$$z = p \cdot f(v_1, v_2) - q_1 v_1 - q_2 v_2 - q_3 \cdot g(v_1, v_2).$$

(b) Another case is a model where the shadow factor depends upon output as well as on the other inputs, for example,

$$x = f(v_1, v_2)$$
$$v_3 = g(x, v_1, v_2). [2]$$

(2b)

---

[1] This type of model, like (1) above, was suggested by PARETO. For references see STIGLER (1946), p. 365.

[2] An empirical example has been given by FRISCH (1935) in what was perhaps the first attempt to derive the production function for an industrial process direct from engineering data. The process in question is the making of a certain type of chocolate. The basic mix (chocolate paste) is composed of a number of ingredients—cocoa beans, sugar, etc.—in fixed proportions according to a given recipe as determined by quality requirements (taste, etc.);

This model can be formally reduced to a case of the type (2a) if the first equation is substituted for x in the second. Alternatively, it may be written in the form

$$x = f(v_1, v_2) = g(v_1, v_2, v_3).\,[1]$$

after heating, the paste is poured into moulds and subsequently cooled. In order to increase the liquidity of the paste such as to reduce the percentage of defective castings, an additional fat ingredient (pure cocoa fat) is mixed into the paste. The resulting increase in cost of raw materials per kg. of chocolate—cocoa fat is more expensive than the basic paste—must be weighed against the reduced cost of remoulding and cooling the defective castings (not against a reduction in the waste of raw materials since the defective castings are reprocessed). This is obviously a substitution problem, the solution of which depends on the relative factor prices.

With a view to geometric illustration in an isoquant diagram, FRISCH analyses this problem in terms of two factors only: additional cost per kg. incurred by adding v kg. of pure cocoa fat to 1 kg. of original paste, and moulding-cooling cost per kg. of finished chocolate. However, the first of these factors cannot be identified with a single homogeneous input. Actually there are (at least) three physical inputs involved in the process, namely the quantity of chocolate paste (v_1, in kg.), the quantity of pure cocoa fat (v_2, also in kg.), and the amount of moulding-cooling work (v_3) incorporated in the finished product; v_3, being a homogeneous complex of labour and other factors entering in fixed proportions, may be measured by the quantity of chocolate (in kg.) that is processed. Further, let x be output of finished chocolate in kg.

Then the production model consists of the following relations. The first is a material balance

(i) $x = v_1 + v_2.$

The second connects v_3 with x. If there had been no defective castings, v_3 would have been a limitational shadow factor proportional to (in the present case, equal to) x. However, some of the castings do not fill the mould completely and are therefore poured back into the paste and remoulded. (Had they gone to waste, (i) would no longer hold.) Therefore, v_3 is greater than x and the relation becomes

$$v_3 \cdot (1 - h) = x$$

where $h\,(<1)$ is the percentage of defective units in each moulding-cooling operation; h is an empirically known decreasing function of the fat content,

$$h = h(v_2/v_1).$$

Substituting in the above relation we have

(ii) $v_3 = \dfrac{x}{1 - h(v_2/v_1)}.$

Eqs. (i)–(ii) constitute a model of the type (2b), subject to which the total cost of the three inputs is to be minimized for given x; the analytical solution is seen to agree with FRISCH's result (op. cit., eq. (7), p. 21).

The practical interpretation of the equilibrium relation (ii) is that the first moulding-cooling operation results in $(v_1 + v_2) \cdot (1 - h) = x \cdot (1 - h)$ kg. of satisfactory chocolate product; $x \cdot h$ kg., being defective, are subsequently reprocessed, resulting in $x \cdot h^2$ kg. of defective castings; and so on until there are no defective castings left. The total amount of moulding-cooling work (quantity of paste processed in the sequence of operations) is

$$v_3 = x + xh + xh^2 + \cdots = \frac{x}{1 - h}$$

and total finished output is

$$x(1 - h) + xh(1 - h) + xh^2(1 - h) + \cdots = x \cdot \frac{1 - h}{1 - h} = x.$$

Thus the equilibrium (ii) is attained in a sequence of operations. (Cp. the interpretation of a multiplier model for the determination of national income.)

[1] This is an example of what FRISCH has called "connected factor rings": the technology is characterized by several simultaneous production functions in wholly or in part the same inputs; cf. FRISCH (1953), pp. 6 f. GEORGESCU-ROEGEN (1935), pp. 48 f., mentions the case $x = f(v_1, v_2, v_3) = g(v_1, v_2, v_3)$ as a possible formal generalization of the production function but questions its practical relevance.

B. Constrained Substitution

Suppose that the inputs of a production process are all substitutable for one another but that the variation is constrained by one or more side relations in some or all of the input variables[1]. A simple example is the model

$$x = f(v_1, v_2, v_3)$$
$$g(v_1, v_2, v_3) = 0. \tag{3}$$

Whereas the previous models (1), (2), (2a), and (2b) were characterized by the number of degrees of freedom being equal to the number of substitutional inputs—for each shadow factor an additional equation was introduced—factor constraints like the second equation of (3) reduce the number of independent input variables in the production function proper, f. (In the extreme case of m inputs and $m-1$ constraints the inputs will be limitational, but the general case of constrained substitution allows for some degree of variability in the factor proportions.)

Models of the type (2b) can be expressed in this form, but (3) may also be an "original" model with an interpretation of its own. For example, although any point (x, v_1, v_2, v_3) satisfying the first equation may be feasible in the sense that the factor combination yields a positive amount of output, some factor combinations must be ruled out because the resulting product does not conform to the quality standard that is required; in this case the side condition $g = 0$ represents a quality constraint defining a region of factor combinations which give a satisfactory product.

Quality constraints may also have the weaker form of inequalities in some or all of the v_i, in which case they merely narrow down the region of substitution without affecting the dimensionality of the production model; the side conditions come in as boundary conditions in the optimization problem[2]. Capacity limitations of the type $v_s \leq \bar{v}_s$ are another special example of factor constraints in inequality form.

Necessary conditions for maximum profit subject to (3) are found by maximizing the Lagrangian

$$L = z + u \cdot g = p \cdot f(v_1, v_2, v_3) - q_1 v_1 - q_2 v_2 - q_3 v_3 + u \cdot g(v_1, v_2, v_3)$$

where u is a Lagrange multiplier. The conditions are

$$(-u =) \frac{p \cdot f_1' - q_1}{g_1'} = \frac{p \cdot f_2' - q_2}{g_2'} = \frac{p \cdot f_3' - q_3}{g_3'}$$

which together with the two equations of (3) determine the optimum values of the four variables.

C. Complementary Groups of Substitutional Inputs

Another possible generalization is a production model of the form

$$x = f(v_1, v_2) = g(v_3, v_4) \tag{4}$$

---

[1] Cf. BORDIN (1944), p. 77.

[2] The problem of quality specifications in the production model is dealt with in some detail in Ch. VIII.

which is characterized by separate factor groups[1]. One interpretation is that the production process is composed of several distinct sub-processes whose production functions—in the present case, f and g—have no inputs in common.

The inputs belonging to each group—for example, v_1 and v_2—are substitutable for one another while the groups as such are complementary in the sense that a higher rate of output requires an increase in both $f(v_1, v_2)$ and $g(v_3, v_4)$; output cannot be kept constant with more of v_1 and v_2 and less of v_3 and v_4. The factor groups are in much the same position with respect to each other and to output as are the several "production functions" in cases of limitationality (perfect complementarity) such as

$$x = f_1(v_1) = f_2(v_2)$$

of which (4) is a generalization[2]. This becomes obvious if model (4) is rewritten in terms of input amounts available:

$$x = \text{Min} \ [\varphi(w_1, w_2), \psi(w_3, w_4)]$$

where

$$\varphi(w_1, w_2) = \text{Max} \ f(v_1, v_2) \quad \text{for} \quad v_1 \leq w_1, \quad v_2 \leq w_2$$

and

$$\psi(w_3, w_4) = \text{Max} \ g(v_3, v_4) \quad \text{for} \quad v_3 \leq w_3, \quad v_4 \leq w_4.$$

The least-cost point for given $x = \bar{x}$, as determined by minimizing the cost expression subject to $f - \bar{x} = 0$, $g - \bar{x} = 0$, is characterized by

$$q_1/f_1' = q_2/f_2'$$

which together with the first side condition determines v_1 and v_2, and

$$q_3/g_3' = q_4/g_4'$$

which must be combined with $g(v_3, v_4) = \bar{x}$ to give the inputs of the second factor group. In other words, the familiar conditions for minimum cost under continuous substitution apply to each factor group separately; the least-cost combination of v_1 and v_2 is independent of the prices of v_3 and v_4, and vice versa.

For variable output, maximization of the Lagrangian expression

$$L = px - q_1v_1 - \cdots - q_4v_4 + u_1 \cdot [f(v_1, v_2) - x] + u_2 \cdot [g(v_3, v_4) - x]$$

---

[1] The case was mentioned by PARETO, although in the form

$$f_1(a_t, a_v, \ldots) = 0, \qquad f_2(a_u, a_x, \ldots) = 0$$

where a_t is the production coefficient of factor no. t:

$$a_t = v_t/x$$

and so on. (Still other inputs are assumed by PARETO to be constant in the general case.) Cf. PARETO (1897), § 714, p. 83n.; see also SCHULTZ (1929), pp. 549—551. In the form used here—cf. (4)—the model has been treated by GEORGESCU-ROEGEN (1935), p. 48, and also by FRISCH (1953), pp. 6 f. (under the name "disconnected factor rings"); cf. also FRISCH (1932), p. 64.

[2] FRISCH (1953), p. 6, cites the example of a compound product whose parts are manufactured separately. When the parts are not made in equal numbers ($f \neq g$), the production function for the composite product is $x = \text{Min} \ [f, g]$, the level of output being determined by the bottleneck just as in the case of limitationality. Cf. Ch. IX, C below.

with respect to x and $v_1, \ldots, v_4$ leads to the following conditions for maximum profit:

$$p = u_1 + u_2, \quad u_1 = q_1/f_1' = q_2/f_2', \quad u_2 = q_3/g_3' = q_4/g_4'$$

where u_1 and u_2 are Lagrange multipliers; $u_1 + u_2$ is to be interpreted as marginal cost at the optimal point since

$$\frac{dc}{dx} = \frac{q_1 dv_1 + \ldots + q_4 dv_4}{dx} = u_1 \cdot \frac{f_1' dv_1 + f_2' dv_2}{dx} + u_2 \cdot \frac{g_3' dv_3 + g_4' dv_4}{dx} = u_1 + u_2,$$

and u_1 and u_2 are the specific marginal costs associated with the respective factor groups (sub-processes).

Chapter VI

The General Single-Product Model

(a) In the previous chapters (III—V) a number of different production functions have been discussed. The cases presented do not, however, form an exhaustive catalogue of single-output production models: other cases are possible and the characteristic features of the respective models discussed may be combined in various ways. Only empirical research can bring to light which of the models are of practical relevance and which are not.

From a formal—perhaps somewhat formalistic—point of view, the various cases may be thought of as special cases of a more general model of production. One way of writing all the cases in a common form is to express output as a function of inputs available,

$$x = \varphi(w_1, w_2, \ldots, w_m) \tag{1}$$

where the w_i are all independent variables. The various cases can then be classified according to the form of the function φ. In order to do this, however, it is necessary to know the more fundamental technological relationships—for example, technically given coefficients of production, factor constraints, etc.—which characterize the process in question. Engineering information of this kind will usually be given in terms of input amounts consumed, v_i. For this reason—and for other reasons already stated—the formulation (1), though quite legitimate, is scarcely adequate to our purpose.

(b) In terms of the variables v_i, not all independent in the general case, most of the models presented[1] can be written in the multi-equation form

$$x = f_k(v_1, v_2, \ldots, v_m) \qquad (k = 1, 2, \ldots, M \leq m) \tag{2}$$

where the number of degrees of freedom (independent variables), $N = m + 1 - M$, ranges from 1 to m.

The number of equations, M, relative to the number of inputs, m, is in a sense the most important characteristic of any particular case since the dimensionality of a production model reflects the possibilities of factor substitution: a model in which all inputs are independent and continuously substitutable is the special case $M = 1$ (i. e., $N = m$) and the opposite extreme is that of $M = m$ (i. e., $N = 1$) where all inputs are limitational and no substitution is possible. The various models of Ch. V represent intermediate cases in the sense that $1 < M < m$.

---

[1] With the sole exception of the linear discontinuous-substitution model (cf. Ch. III, B).

For purposes of classification and interpretation it is also important to note which variables occur in which equations. Most of the various particular models that can be written in the common form (2) are not regular in the sense that all equations contain all of the variables, each particular type of model being distinguished by a special pattern of zero coefficients; certain inputs, for example, may be in a special position in that they are uniquely related to output or to certain inputs, and so forth.

(c) Alternative formulations of the input-output relationships can be derived from the general model (2). For example, eliminating x from all but one of the equations we get a model consisting of a single production function with $M - 1$ side conditions in the inputs:

$$x = f(v_1, v_2, \ldots, v_m)$$
$$g_k(v_1, v_2, \ldots, v_m) = 0 \qquad (k = 1, 2, \ldots, M - 1) \tag{2a}$$

which is formally a case of constrained substitution in m inputs[1].

Since model (2) or (2a) has N degrees of freedom, we can choose N independent input variables—say, $v_1, v_2, \ldots, v_N$—and solve the equations for output and the remaining $m - N$ inputs as dependent variables[2]:

$$x = F(v_1, v_2, \ldots, v_N)$$
$$v_j = G_j(v_1, v_2, \ldots, v_N) \qquad (j = N + 1, \ldots, m). \tag{2b}$$

In this equivalent form the general model can be formally analysed as a case of continuous substitution in N independent inputs with $m - N$ shadow factors[3] [4]. Model (2b) is particularly well suited to the solving of optimiza-

---

[1] For example, the model $x = f_1(v_1, v_2, v_3) = f_2(v_1, v_2, v_3)$ becomes

$$x = f_1(v_1, v_2, v_3)$$
$$g(v_1, v_2, v_3) = 0$$

where $g = f_1 - f_2$. The limitational model $v_1 = a_1 x$, $v_2 = a_2 x$ can be written

$$x = v_1/a_1 = v_2/a_2$$

from which, by addition and subtraction respectively, we get

$$x = (1/2a_1) \cdot v_1 + (1/2a_2) \cdot v_2$$
$$(1/a_1) \cdot v_1 - (1/a_2) \cdot v_2 = 0.$$

[2] Cf. FRISCH (1953), pp. 22 ff., SHEPHARD (1953), pp. 30 ff., and RICOSSA (1955).

[3] FRISCH (1953), pp. 22 ff.

[4] For example, the model

$$v_1 = a_1 x, \qquad v_2 = a_2 x$$

(where $N = 1$) becomes

$$x = (1/a_1) \cdot v_1, \qquad v_2 = (a_2/a_1) \cdot v_1$$

with v_1 as the independent variable. The case of complementary input groups

$$x = f(v_1, v_2) = g(v_3, v_4)$$

($N = 3$) gives, adding and subtracting the equations respectively,

$$2x = f + g$$
$$0 = f - g;$$

solving the latter equation for, say, v_4 and substituting in the former we get x as well as v_4 expressed as functions of v_1, v_2, and v_3.

In general the N independent variables cannot be picked arbitrarily from among the total set of m inputs. For example, in the case

$$x = f(v_1, v_2), \qquad v_3 = a_3 x, \qquad v_4 = a_4 x$$

($N = 2$), v_3 and v_4 will not do because they are both proportional to x.

tion problems: it can be used directly to eliminate dependent variables in the cost or profit expression, thus reducing the number of side conditions in the maximization problem[1]. Substituting the functions G_j in the cost function,

$$c = \sum_{i=1}^{N} q_i v_i + \sum_{j=N+1}^{m} q_j G_j(v_1, \ldots, v_N),$$

$m - N$ variables and equations are got rid of and the least-cost point for given $x = \bar{x}$ is determined by minimizing this expression subject only to the side condition $\bar{x} = F$—just as in ordinary cases of continuous factor substitution, only the necessary conditions for minimum cost now have the form

$$(u =) \frac{q_1 + \sum_j q_j \cdot \delta G_j / \delta v_1}{F_1'} = \cdots = \frac{q_N + \sum_j q_j \cdot \delta G_j / \delta v_N}{F_N'}$$

where u is a Lagrange multiplier to be interpreted as marginal cost. Profit maximization is even simpler because there are no side equations left when x has been replaced by the function F.

(d) Another formulation of the production relations is the parametric production model[2]

$$\begin{aligned} x &= \Phi(t_1, t_2, \ldots, t_N) \\ v_i &= \Psi_i(t_1, t_2, \ldots, t_N) \qquad (i = 1, 2, \ldots, m) \end{aligned} \tag{3}$$

where $t_1, t_2, \ldots, t_N$ are independent parameters. It is always possible to write (2) in this form—(2b) can be thought of as a parametric model with $v_1, v_2, \ldots, v_N$ as parameters—but (3) may also have an interpretation of its own. Some or all of the parameters may represent "engineering variables"[3] which cannot be identified with economic variables (inputs), such as machine speeds, hours of operation per period, etc.[4]. Another example is the linear model with discontinuous substitution, where Φ and the Ψ_i are linear functions of the activity levels λ_k. In special cases of this kind—when $N \leq m$—model (3) can be written in the form (2b), using N of the equations $v_i = \Psi_i$ to eliminate the parameters[5], but when there are "too many" parameters in (3) this is no longer possible. (3), therefore, is a more general formulation of the production model; all the models of Chs. III–V can be expressed in this form.

The optimal allocation in production, then, is that configuration of x, the v_i, and $t_1, \ldots, t_N$ which represents a maximum of the profit function subject to (3) and subject to additional constraints in the form of inequalities—non-negativity requirements, capacity limitations, quality specifications, etc.—which do not reduce the number of degrees of freedom in the production model but which may well affect the position of the optimal point.

---

[1] Cf. Appendix 1, A.
[2] Cf. FRISCH (1953), pp. 45 ff.
[3] Cf. CHENERY (1953), especially p. 302.
[4] Examples are given in Ch. VII.
[5] Cf. the numerical example in Ch. III, B above.

Chapter VII

Divisibility, Returns to Scale, and the Shape of the Cost Function

A. Fixed Factors and the Production Function

The neoclassical theory of production traditionally assumes, in short-run analysis of the individual firm (or process), that not all of the factors of production involved are variable. Those which represent the plant are assumed to be fixed in an absolute sense, the plant being composed of given indivisible units of capital equipment.

In a two-factor analysis of an agricultural production process let v_1 denote the factor "land" while v_2 is the variable factor (e.g. labour, or a homogeneous complex of variable inputs), and let the production function be

$$x = x(v_1, v_2). \tag{1}$$

(Alternatively, (1) may be interpreted as an industrial production function where v_1 represents the fixed equipment and v_2 represents a complex of variable inputs.) The short-run situation is characterized by v_1 being fixed, $v_1 = \bar{v}_1$. Then output may as well be written as a function of v_2 only,

$$x = x(\bar{v}_1, v_2) = f(v_2), \tag{2}$$

where the value of v_1 is no longer specified in the production function but is implicit in the specification of the function f [1]. For variations in v_2, output is assumed to follow the law of variable proportions as shown in *Fig. 34a (OABC)* [2]. The corresponding total cost function (including fixed costs) will have the shape $DEFG$ indicated in *Fig. 34b*, the cost of the variable factor being its price times the inverse of the production function:

$$c = q_2 v_2 = q_2 f^{-1}(x).$$

The marginal cost curve, $c' = q_2/f'$, will have the familiar U-shape which is so characteristic of neoclassical theory. The curvilinear shape of the pro-

---

[1] In the words of v. STACKELBERG (1938), p. 78: „Bekanntlich wird bei kurzfristiger Änderung der Ausbringung ein Teil der Produktionsmittel in unveränderten Mengen aufgewendet. Die Aufwandsmengen dieser Produktionsmittel treten . . . in der Produktionsfunktion nicht als Variable in Erscheinung, sondern sind mitbestimmend für die Form dieser Funktion. Man kann sie . . . einfach fortlassen."

[2] The production curve is especially likely to have this particular shape if the production function is homogeneous of degree one in v_1 and v_2 (cf. Ch. IV, B and Appendix 2), an assumption often made in neoclassical theory.

duction and cost curves is due to the fact that variations in the level of output must necessarily be accompanied by changes in the proportion of v_2 to v_1 when the latter cannot be adjusted to fluctuations in x.

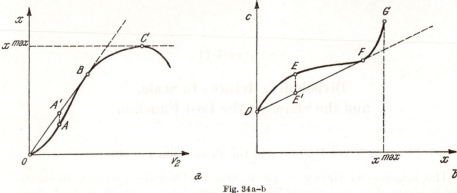

Fig. 34 a–b

Now if we allow for divisibility, the model becomes less rigid in that both inputs will be capable of continuous variation, although in the short-run situation v_1 is available only up to a limit, $\bar{v}_1$. The cost function is now defined along an expansion path determined by minimizing $c = q_2 v_2$ [1] subject to (1) for parametric $x = \bar{x}$ and to the capacity restriction $v_1 \leq \bar{v}_1$, which has replaced the "built-in" capacity parameter $\bar{v}_1$ in (2). As we have seen[2], the variation of x with v_2 in Fig. 34 a now follows the path $0 A' B C$ [3], which—assuming homogeneity—is linear up to B, the point where the expansion path is deflected because v_1 becomes $= \bar{v}_1$. [4] The divisibility of v_1 makes it possible to rule out points on $0AB$; point A is inefficient since $x_1' < 0$, and output can be increased by reducing the utilization of v_1 (v_2 being the same) until x_1' becomes $= 0$ at point A'. Along $0A'B$ the relative factor proportion is constant and the elasticity of production, ε, is equal to one.

Correspondingly, for efficient production the total cost curve $DEFG$ in Fig. 34 b is replaced by $DE'FG$ where marginal cost is constant along the linear segment $DE'F$. [5]

---

[1] Assuming $q_1 = 0$.

[2] Cf. Ch. IV, E (esp. Fig. 32).

[3] This observation—without reference to the corresponding cost function—was apparently first made by GLOERFELT-TARP (1937), pp. 225—233, and later by CHIPMAN (1953), p. 106. The inefficiency of points on $0AB$ had previously been pointed out by KNIGHT (1921), pp. 100 f.

[4] It does *not* follow that the marginal productivity of the variable input v_2 will be constant up to point B; x_2' is defined as $\delta x / \delta v_2$ for v_1 constant, i.e., as the slope of the tangent at a point on the production curve $0ABC$. All that the argument states is that the expansion path in Fig. 34 a will not coincide with $0ABC$.

[5] Cost functions of this kind have been treated in the literature but usually without precise indication of how they may come about. SCHNEIDER (1934), pp. 13 and 49, first justified them on purely empirical grounds, but in his later works—cf. SCHNEIDER (1958), pp. 170 and 189 ff.—he has derived a cost function of this shape from a parametric production model (cf. below).

The crucial question, then, is whether or not it is realistic to assume that the fixed factor is divisible—a question which is closely bound up with that of homogeneity, the usual common-sense argument for homogeneity in production functions being that twice the amount of *all* inputs will double output because it is a mere repetition (in time or space)[1].

This is ultimately a matter of empirical research, not a question to which the pure theory of production and cost can readily provide a general answer. Still, it would seem as though the problem of divisibility has traditionally been given rather less than due attention in the literature. In the following we shall deal with plant divisibility in some detail to see in what way the fixed factor can be divisible and how divisibility affects the cost functions belonging to a variety of hypothetical production models.

B. The Dimensions of Capacity Utilization

1. Spatial Divisibility

(a) Let us first assume that the fixed factor is *physically divisible in space* but indivisible in the time dimension[2].

An obvious example is an agricultural production process where several variable inputs (labour, fertilizer, etc.) cooperate with the services of a plot of land during a period of a year. The factor "land" is physically divisible in that it is not necessary to cultivate the entire available acreage; actually factor combinations involving negative marginal productivity of land will be ruled out as inefficient. Land, however, is not divisible in the time dimension in the sense that it is possible to use it for part of the production period only. This is due to the fact that grain-growing is an example of *intermittent* production: the product emerges not continuously during the production period but at discrete points of time—once a year—whereas the fixed factor must be employed continuously throughout the period. (For this reason it does not matter whether the input of land is defined as the acreage as such or as its services, e.g. in terms of acre-years, as long as we consider a given period of a year.)

Another example[3] is an industrial plant composed of a number of identical units (machines, reactors, etc.), some of which may not be utilized, but which are indivisible in the time dimension in that those units which are actually used have to be effectively operated during the entire period considered[4]; only in this case the fixed factor is not continuously variable since an integral number of units of equipment will have to be operated.

---

[1] In Fig. 34a, for example, if A' is halfway between 0 and B, the expansion from 0 to A' is identically repeated in moving on from A' to B.

[2] In the following, only such fixed factors are considered as are directly productive (cf. Ch. II above). There is no need to include buildings in the list of specified inputs since their productiveness consists in mere presence.

[3] DEAN (1941), pp. 4 ff.

[4] Blast furnaces, for example, must be operated continuously without breaks, cf. DEAN (1951), p. 275.

Consider a given plant composed of N identical units of a fixed factor, for example, N machines or N acres of land. Each of these units produces z units of product per period, using the quantities $u_2, u_3, \ldots, u_m$ of $m-1$ variable inputs (and, by definition, $u_1 = 1$ unit of the fixed factor). Let n of the N available units of equipment (or land) be in use during the whole period considered. We can then express plant output x and the total quantities of inputs $v_1, v_2, \ldots, v_m$ per period as follow:

$$x = n \cdot z$$
$$v_1 = n \cdot u_1 = n \quad (\leq N) \tag{3}$$
$$v_i = n \cdot u_i \quad (i = 2, 3, \ldots, m).$$

These equations together with the technological relation or relations connecting z with the u's—i.e., the production function for the individual unit of equipment—constitute a parametric production model for the whole plant; eliminating the parameters n, z, and the u's we get the production function in terms of x and the v's only.

(b) For example, let all inputs—including the fixed factor—be *limitational* and proportional to output. In this case, since $u_1 = 1$ by assumption, output per machine per period and the variable inputs $u_2, u_3, \ldots, u_m$ are all constant:

$$z = \bar{z}, \quad u_i = \bar{u}_i \quad (i = 2, 3, \ldots, m). \tag{4a}$$

Substituting (4a) in (3) we get a parametric production model which reduces to a model of the form

$$v_i = a_i \cdot x \quad (i = 1, 2, \ldots, m) \tag{5a}$$

where

$$a_1 = \frac{1}{\bar{z}}, \quad a_2 = \frac{\bar{u}_2}{\bar{z}}, \ldots, a_m = \frac{\bar{u}_m}{\bar{z}}$$

are constant coefficients of production. The only kind of variation permitted by the model consists in varying n, the number of units of equipment employed (i.e., the number of identical processes), which cannot exceed N so that the capacity limitation is

$$v_1 \leq N \text{ or } x \leq \frac{N}{a_1}.$$

The expansion path in this case is a straight line in m-dimensional factor space (up to capacity) and total variable cost is a linear function of the level of output,

$$c = \sum_{i=2}^{m} q_i v_i = \bar{c} \cdot x$$

where $\bar{c} = \sum_{i=2}^{m} q_i a_i$ is the constant marginal cost (cf. Fig. 1 a–b)[1].

---

[1] Strictly speaking, when n is not continuously divisible but can assume integral values only ($n = 1, 2, \ldots, N$ machines), the expansion path as well as the cost function will be a sequence of discrete points of the straight line. Marginal cost is no longer defined, but incre-

(c) On the other hand, if $u_2, u_3, \ldots, u_m$ are continuously *substitutable* inputs (u_1 still being $= 1$ by definition), we have

$$z = \varphi(u_2, u_3, \ldots, u_m). \tag{4b}$$

Eliminating the $m + 1$ parameters n, z, and $u_2, u_3, \ldots, u_m$ from the parametric model (3)–(4b) we get the production function for the whole plant:

$$x = v_1 \cdot \varphi\left(\frac{v_2}{v_1}, \frac{v_3}{v_1}, \ldots, \frac{v_m}{v_1}\right) = x(v_1, v_2, \ldots, v_m) \tag{5b}$$

where we now have m substitutable inputs of which v_1 represents the degree of utilization of the fixed equipment—i.e., the "extensive" dimension of capacity utilization—whereas an increase in output brought about by increasing $v_2, \ldots, v_m$ for v_1 constant is to be interpreted as more intensive utilization of the plant (intensive farming, etc.).

The function (5b) is seen to be homogeneous of the first degree, hence the expansion path is a straight line in m-dimensional factor space as long as $n < N$ and the corresponding phase of the cost function will be linear. When the capacity is fully utilized in the extensive dimension ($n = N$), the expansion path is deflected and further expansion requires more intensive operation, increasing $v_2, \ldots, v_m$ and thus changing the factor proportions. Because v_1 is now fixed, output cannot be expanded indefinitely; the second phase of the expansion will be characterized by decreasing returns to scale and marginal cost will be rising up to the point of maximum capacity[1]. Thus the cost function will have the same general shape as that shown in Fig. 32[2].

---

mental cost $\Delta c / \Delta x$ (for $\Delta x = \bar{z} = 1/a_1$) is the same at all points and equal to average variable cost. The stepwise expansion path derived by BREMS (1952a, b) for this case is based on the assumption that the isoquants are L-shaped, including points characterized by less than full utilization of the available factor quantities. Even if this definition of the production function is adopted, such points cannot belong to the expansion path because they are clearly inefficient, cf. Ch. III, A.

[1] Cf. Ch. IV, E.

[2] When n is a discrete variable assuming integral values only, the expansion path will be somewhat different; in the two-factor case the graphical picture will be as follows:

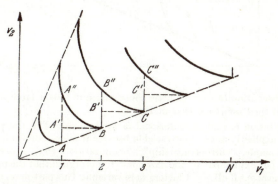

where the expansion path is composed of the disconnected line segments AA', BB', etc.; points on $A'A''$, $B'B''$, etc. are not efficient since the points B, C, etc. represent larger output

(d) The linear shape of the cost curve within the limit $v_1 \leq N$ is derived from the assumption of homogeneity, which in turn is related to that of plant divisibility. When the fixed factor is divisible, it is tempting to make the further assumption that the production process is also divisible in the sense that, as output is expanded by more units of the fixed factor coming into use, each consecutive unit cooperates with the same quantities of the other inputs to yield the same amount of output. This assumption is implicit in the model, where (3) states that total output and total inputs are the same multiple of output and inputs associated with a unit of the fixed factor, the latter variables being related by (4a) or (4b), so that output can be expanded by identical repetition in the space dimension (more extensive cultivation of the soil, utilizing more machines, etc.).

It is precisely this element of repetitiveness, based on factor divisibility, that underlies the usual common-sense justification of constant returns to scale. The argument is plausible enough in many cases; a more general theoretical model, however, must allow for the possibility of economies (or diseconomies) of scale within the plant. Analytically this means that n may appear as an additional variable in (4b):

$$z = \varphi(u_2, u_3, \ldots, u_m, n);$$

(5b) then becomes

$$x = v_1 \cdot \varphi\left(\frac{v_2}{v_1}, \frac{v_3}{v_1}, \ldots, \frac{v_m}{v_1}, v_1\right) \qquad (6)$$

which is no longer homogeneous and which therefore leads to a non-linear expansion path and a non-linear cost function[1].

---

at less cost so that it will pay to produce these larger outputs and dispose of the amount of output not required. The corresponding cost function will also be discontinuous:

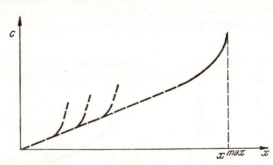

However, these discontinuities are of little consequence if N is a large number; the shape of the curves will then approach that of Fig. 32.

[1] One interpretation is that an additional factor is involved in the production process but not specified explicitly as an input variable because it is an indivisible fixed factor (representing, for example, "entrepreneurship", "organization", etc.), $v_0 = \bar{v}_0$. The absence of linearity can then be "explained" by the fixity of v^0 if it is argued that the production function ought to be homogeneous if all $m+1$ factors were variable. This postulate, however, is empirically meaningful only if v^0 can be defined in such a way as to be physically specifiable in quantitative terms, i.e., measurable (cf. Appendix 2).

2. The Time Dimension

(a) Next let us consider the case where the fixed factor is physically indivisible but *divisible in the time dimension* in that it is possible to vary the number of hours of operation per period of calendar time. For example, the utilization of a machine in terms of machine hours per month may be variable within the limit set by the total number of hours the machine is available in a month. In such cases it is the *services* of the fixed factor which are represented by the input variable v_1, not the factor itself (e.g. machine hours rather than machines).

Clearly, full divisibility of this kind is possible only when the production process considered is a *continuous* one in the sense that output and inputs are continuous flows in time. In the case of intermittent production, as we have seen above, the process and thus also the use of the fixed factor in a period is an indivisible whole.

In much of the literature on the theory of production the time dimension is eliminated by the assumption that the plant is operated at full capacity in time, for example, that the number of effective working hours per day is fixed[1], so that output per period (day) can be varied only by more or less intensive operation. While this may in fact be correct in some cases of continuous production[2], as a general assumption it is unnecessarily restrictive and leads to factor combinations which are uneconomical or even clearly inefficient.

To clarify the argument, let us consider a hypothetical productive process. A plant produces z units of a single product with labour, energy, raw material, and other variable inputs in quantities $u_2, u_3, \ldots, u_m$ in physical units *per hour of effective operation*[3]. The plant as such is fixed and physically indivisible (e.g. one machine) but its services, as measured in plant hours of effective operation (machine hours), are continuously divisible; the input of plant services per hour of operation, u_1, is of course $= 1$ by definition.

The rate of output *per period of calendar time* (e.g. per day) and inputs per period, when the plant is operated t hours out of the T hours available per period, will be

$$x = t \cdot z$$
$$v_1 = t \cdot u_1 = t \quad (\leq T) \tag{7}$$
$$v_i = t \cdot u_i \quad (i = 2, 3, \ldots, m).$$

When the production function for an hour's operation is known, we can eliminate t, z, and the u's and the production function for output per period emerges.

The introduction of the time factor adds a new dimension to the production model; as the equation

$$x = t \cdot z$$

---

[1] Cf., for example, CARLSON (1939), p. 15 n.

[2] Such as processes which are continuous also in the sense that, for technical reasons, the plant must be operated continuously without breaks (e.g. blast furnaces, cf. above).

[3] Not per *calendar* hour.

shows, the quantity of product per period can be expanded not only by applying larger amounts of variable inputs per hour (provided that substitution is possible), thus increasing the hourly rate of output by more intensive utilization of the fixed factor, but also by increasing the number of hours of operation per period, that is, by more extensive operation in the time dimension. (The capacity of the plant, similarly, has two dimensions, capacity per period being maximum output per hour times the number of hours (T) the plant is available per period.) And a given level of output can be attained by different combinations of t and z, producing more per hour in fewer hours per period; in so far as they represent different factor combinations, a new dimension has been added to the substitution possibilities.

(b) Assume first that all inputs are *limitational* with constant input-output coefficients. Then, by analogy with (4a) and (5a) above, output and variable inputs per hour of effective operation will be constant since the input of machine services per hour is constant by definition ($u_1 = 1$):

$$z = \bar{z}, \quad u_i = \bar{u}_i \qquad (i = 2, 3, \ldots, m). \tag{8a}$$

The parametric production model (7)–(8a) reduces to

$$v_i = a_i \cdot x \qquad (i = 1, 2, \ldots, m) \tag{9a}$$

where

$$a_1 = \frac{1}{\bar{z}}, \quad a_i = \frac{\bar{u}_i}{\bar{z}} \qquad (i = 2, 3, \ldots, m)$$

are constant coefficients of production. No factor substitution for x constant is possible. The only kind of variation permitted by this model is a proportional expansion of output and inputs in the time dimension up to the limit $t = T$; in terms of output per period the capacity limitation is

$$x \leq \frac{T}{a_1}.$$

The expansion path is linear and so is the cost function.

(c) Suppose instead that the u's are continuously *substitutable* factors. Then we have

$$z = \varphi(u_2, u_3, \ldots, u_m) \tag{8b}$$

and this combines with (7) to give

$$x = v_1 \cdot \varphi\left(\frac{v_2}{v_1}, \frac{v_3}{v_1}, \ldots, \frac{v_m}{v_1}\right) \tag{9b}$$

which is perfectly analogous to (5b) as derived from (3)–(4b). The time dimension, as represented by $t = v_1$, adds a new variable and substitutable factor to the $(m-1)$ already present in (8b); the latter equation, in which the input of plant services per hour is fixed ($u_1 = 1$), represents a more restricted field of variation, presumably obeying the law of variable proportions. When all m factors are free to vary, i.e., for $v_1 = t < T$, the expansion path in m-dimensional factor space will be a straight line whose direction depends on the factor prices $q_2, q_3, \ldots, q_m$ (x_1' being $= 0$ for

$q_1 = 0$) and along which marginal cost is constant. When the capacity is fully utilized in the time dimension ($v_1 = T$), further expansion requires more "intensive" operation; this means changing the factor proportions as z increases—cf. (8 b)—so that we get diminishing returns and rising marginal cost at this stage. In other words, the cost function will have the shape of Fig. 32.

If the inputs are substitutable but in a discontinuous fashion because the production function consists of a finite number of linear activities, the curvilinear phase of the cost curve will be replaced by a broken line pieced together from linear segments; marginal cost, accordingly, will be a step function (cf. Fig. 10).

(d) Another class of models, presumably relevant to many industrial production processes, is represented by the "mixed" cases where not all of the inputs are independent variables; some of them are mutual substitutes whereas others are shadow factors, either limitational factors or tied up with the substitutional inputs by additional relations. Let $u_1, \ldots, u_4$ be inputs of machine hours, man-hours, energy (or fuel), and raw material, all per hour of effective operation, and let us assume throughout that raw material is a limitational input, a units of material being required per unit of output produced. A couple of special cases of the mixed type may then be distinguished.

Let us first assume that the ratio of man-hours to machine hours is fixed, the technical characteristics of the machine being such that it requires a constant number of operators, k, but that the hourly rate of output can be expanded within limits by speeding up the machine, which in turn requires more energy or fuel input per hour. Then, in addition to

$$x = t \cdot z$$
$$v_1 = t \cdot u_1 = t \quad (\leq T) \tag{10}$$
$$v_i = t \cdot u_i \qquad (i = 2, 3, 4)$$

we have

$$z = \varphi(u_3)$$
$$u_2 = k \tag{11a}$$
$$u_4 = a \cdot z.$$

In this case, the only substitution that is possible within the model is that of energy or fuel for the machine-labour factor complex. Eliminating t, z, and the u's from the parametric production model (10)–(11 a) we are left with three equations,

$$x = v_1 \cdot \varphi\left(\frac{v_3}{v_1}\right)$$
$$v_2 = k \cdot v_1$$
$$v_4 = a \cdot x$$

which may also be written

$$x = \frac{1}{k} \cdot v_2 \cdot \varphi \left(k \cdot \frac{v_3}{v_2} \right) = f(v_2, v_3)$$

$$v_1 = \frac{1}{k} \cdot v_2 \quad (\leq T) \tag{12a}$$

$$v_4 = a \cdot x,$$

so that we have a production function in two substitutable inputs, v_2 and v_3, whereas v_1 and v_4 are shadow factors (v_1 being a "factor shadow" and v_4 a "product shadow")[1]. Output per day can be expanded either in the time dimension as represented by v_2 (or v_1), i.e., by working more hours per day, or by applying more fuel or energy per hour during a working day of given length, i.e., by speeding up the machine[2].

Since the production function (12a) is homogeneous of degree one, the expansion path is a straight line in the (v_2, v_3) plane for $t < T$, i.e., for $v_2 < k \cdot T$, and marginal cost

$$c' = \frac{q_2}{f_2'} + q_4 a = \frac{q_3}{f_3'} + q_4 a \ ^3$$

is constant in the region, cf. OC in *Fig. 35*. Expansion along this path represents more extensive utilization of machine capacity in the time dimension: hourly output z and the rates of inputs per hour (the u_t) are constant for proportionate variation in v_2 and v_3. At the point C where t becomes $= T$, capacity is exhausted in the time dimension and further expansion of output will have to follow the path

$$v_2 = T \cdot u_2 = k \cdot T$$

---

[1] Cf. Ch. V, A.

[2] Using a model based on technological assumptions similar to those underlying model (12a)—which he considers typical of industrial production processes—GUTENBERG (1963), pp. 210 ff., has argued that processes of this kind are characterized by the absence of factor substitution. Speeding up the machine requires not only more fuel and energy as well as more raw material; the operators will have to work harder and the heavier demands made on the machine will cause greater wear and tear and higher costs of repair. Hence all inputs are limitational: z cannot be increased for constant inputs of labour and machine service, each of the four inputs being uniquely determined by output.

The answer to this is that the constancy of u_1 and u_2 in (11a) is a consequence of the units of measurement chosen. When labour is measured in man-hours, u_2 remains constant for variations in z and the fact that the operators have to work harder when the machine runs at greater speed is represented by an increase in the (average) productivity of labour,

$$x/v_2 = z/u_2 = \varphi(u_3)/k.$$

To interpret this as an increase in the input of labour is to measure labour in terms of the output it produces (i.e., in terms of man-hours times output per man-hour, $v_2 \cdot (x/v_2) = x$), which is unreasonable not only because the production function is bypassed but also because a man-hour is the unit with which the factor price q_2 is associated (assuming that the workers are paid by the hour).

As to machine services, increased wear and tear and cost of repair per hour can be taken into account by treating them as separate inputs (shadow factors) depending on z along with fuel and raw material, thus purging the factor "machine service" of all other dimensions than the utilization rate in the time dimension.

For a more detailed discussion of Gutenberg's argument see DANØ (1965).

[3] Cf. Ch. V, A.

(CD in Fig. 35) so that we have

$$x = f(kT, v_3) = T \cdot \varphi(v_3/T) \quad \text{or} \quad z = x/T = \varphi(u_3),$$

that is, hourly output is expanded in the intensive dimension. Marginal cost

$$c' = \frac{q_3}{f_3'} + q_4 a \quad \left(= \frac{q_3}{\varphi'} + q_4 a \right)$$

is increasing during this phase (assuming that the function φ is characterized by diminishing returns) and the composite marginal cost function is contin-

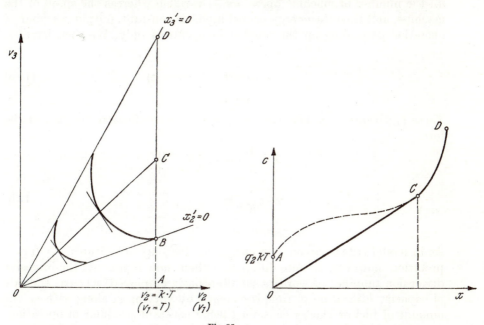

Fig. 35

uous at the point of deflexion, C. Thus the curve of total variable cost has the same shape as that shown in Fig. 32 [1]. The expansion path differs from that of Fig. 32 only in that the linear phase does not follow the boundary of the region of substitution because a positive price is associated with the fixed factor's services, v_2 being proportional to v_1. [2]

Had we assumed in advance that the capacity would be fully utilized in the time dimension, the expansion path would have been $ABCD$ and the marginal cost curve would have been U-shaped. Such an assumption

---

[1] Schneider (1958), pp. 170 and 189 ff., was the first to show explicitly that a cost function of this type can be due to a production model of the kind described here (leaving out the shadow input of raw material). Schneider's treatment differs from the above only in that he does not derive model (12a) but derives the cost function directly from the parametric production model (10)–(11a), indicating the expansion path in a (z, t) diagram. Cf. Ch. IV, B.

[2] In general, the existence of shadow factors proportional to the utilization of the fixed factor is an important reason why the services of the fixed factor cannot always be considered free.

is unnecessarily restrictive in the case of continuous production with a fixed factor that is divisible in time, ignoring as it does an important aspect of substitution, and leads to a cost curve which cannot claim to represent minimum cost for any given level of output per period. Points on the line segment AB represent inefficient factor combinations ($x_2' < 0$) and points on BC, though not technically inefficient, do not correspond to least-cost combinations for $q_2 > 0$.

(e) A formally similar model can be set up to represent the case where u_2, the number of machine operators, is variable whereas the speed of the machine, and thus the energy or fuel input, is constant, output per hour of operation depending on the number of operators only. We then have

$$z = \psi(u_2)$$

$$u_3 = K \quad \text{(constant)} \tag{11b}$$

$$u_4 = a \cdot z.$$

These equations, together with (10), lead to a production function of the form

$$x = \frac{1}{K} \cdot v_3 \cdot \psi\left(K \cdot \frac{v_2}{v_3}\right)$$

$$v_1 = \frac{1}{K} \cdot v_3 \tag{12b}$$

$$v_4 = a \cdot x.$$

As in model (12a), labour and energy (or fuel) are substitutes; the interpretation, however, is quite different. Labour now represents the intensive dimension (number of workers simultaneously employed) whereas the rate of capacity utilization in time is changed by varying v_1 along with v_3, the amount of fuel or energy consumed being constant per hour of operation; output per hour can be increased by more labour-intensive operation, not by increasing the speed of the machine. The model is still homogeneous of degree one and the expansion path as well as the cost function will have the same general shape as in the previous example[1].

(f) Still another variant is a model allowing for variable number of operators as well as for variable machine speed. This implies that u_2 and u_3 are mutually substitutable factors in the production function for hourly output[2], whereas the time dimension is now represented by v_1 only. Combining the assumptions

$$z = \mu(u_2, u_3)$$

$$u_4 = a \cdot z. \tag{11c}$$

with (10) we have

---

[1] Except that u_2 assumes integral values only so that $v_2 = t \cdot u_2$—continuously divisible in the time dimension for $t < T$—becomes a discrete variable for given $t = T$.

[2] The interpretation is that output per hour depends on machine speed (in r.p.m.) and on the number of operators, machine speed in turn being a function of energy or fuel input per hour.

$$x = v_1 \cdot \mu \left(\frac{v_2}{v_1}, \frac{v_3}{v_1} \right)$$

$$v_4 = a \cdot x,$$

(12 c)

a model similar to those above except that there are now three substitutable factors. The expansion path is a straight line in the three-dimensional factor space[1] up to the point $t = v_1 = T$, beyond which further expansion will be accompanied by changing factor proportions and, hence, by rising marginal cost up to the point of maximum capacity (where the partial derivatives of the function μ are zero).

(g) The linear segment of the expansion path and the corresponding linear segment of the cost curve in the examples above are due to the assumed homogeneity of the production function, which in turn reflects the assumption that an hour's operation is a mere duplication of the preceding hour. This is implicit in the assumption that output per hour, $z = x/t$, is independent of t. Suppose, however, that average output per hour depends not only on the factor combination employed but also on the length of the working day (labour productivity declines when the workers get tired, etc.). In model (11 a), for example, this would imply replacing the first equation by

$$z = \varphi(u_3, t),$$

t being here the number of hours of operation per day[2][3]. The first equation of (12 a) would then have the form

$$x = v_1 \cdot \varphi \left(\frac{v_3}{v_1}, v_1 \right) \text{ or } x = \frac{1}{k} \cdot v_2 \cdot \varphi \left(k \cdot \frac{v_3}{v_2}, \frac{1}{k} \cdot v_2 \right)$$

which is not homogeneous of degree one (cf. (6)). The expansion path for minimum cost will no longer be a straight line for $t < T$, nor will the corresponding segment of the cost curve be linear.

Another assumption implicit in the models above, and in the usual derivation of the cost function, is the constancy of factor prices, especially the wage rate. If the capacity limitation in the time dimension allows for overtime or multiple-shift operation, labour cost per man-hour will rise with t in a discontinuous manner and marginal cost will be constant only as long as t is kept within normal hours.

---

[1] Since partial variation of v_1 no longer affects cost—as it did above where v_2 or v_3 was a shadow factor—the expansion path will be on the boundary of the region of substitution ($x_1' = 0$).

[2] If the period considered had been a week, a month, etc. rather than a day, z would not be affected by all kinds of variation in t; for example, it would be possible within limits to vary the number of days of operation per month for constant number of hours worked per day.

[3] One interpretation of the equation is that, although working speed is constant for u_3 constant, being determined by the speed of the machine, the percentage of defective units of product is larger when the operators are tired. (It is also likely to increase with the speed of the machine, i.e., with u_3 for constant t, but this effect is implicit in the shape of the function f when z is interpreted as number of non-defective units made per hour.)

3. Divisibility in Space and Time

In some cases the fixed factor is *divisible in space as well as in time.* By way of example, consider a continuous process which differs from that underlying model (12a) only in that the plant is composed of N identical machines some of which may not be used during the period considered. Let s denote output per machine per hour of effective operation, and let s be a function of fuel input per machine per hour, y_3; each machine is worked by a fixed number of operators, $y_2 = k$. Then total output and total inputs per hour when n machines are operated $(n \leq N)$ are

$$z = n \cdot s = n \cdot \varphi(y_3)$$

$$u_1 = n \qquad\qquad \text{(machine hours per hour} =$$
$$\text{number of machines used)}$$

$$u_2 = n \cdot k \qquad\qquad \text{(man-hours per hour} =$$
$$\text{number of operators)}$$

$$u_3 = n \cdot y_3$$

$$u_4 = n \cdot a \cdot s = a \cdot z.$$

The parametric production model for total output and total inputs per period when the plant is operated t hours $(t \leq T)$ becomes

$$x = t \cdot z = t \cdot n \cdot s = t \cdot n \cdot \varphi(y_3)$$

$$v_1 = t \cdot u_1 = t \cdot n$$

$$v_2 = t \cdot u_2 = t \cdot n \cdot k$$

$$v_3 = t \cdot u_3 = t \cdot n \cdot y_3$$

$$v_4 = t \cdot u_4 = t \cdot n \cdot a \cdot s = a \cdot x$$

which reduces to the three-equation model

$$x = v_1 \cdot \varphi\left(\frac{v_3}{v_1}\right) \quad \text{or alternatively} \quad x = \frac{1}{k} \cdot v_2 \cdot \varphi\left(k \cdot \frac{v_3}{v_2}\right)$$

$$v_2 = k \cdot v_1 \tag{13}$$

$$v_4 = a \cdot x.$$

This model is formally analogous to (12a); the only difference in interpretation is that the input of plant services (machine hours)—with its shadow, labour input—is variable in two extensive dimensions, v_1 being number of machines operated times number of operating hours per period. The variation is subject to the restrictions $t \leq T$, $n \leq N$, i.e., $v_1 \leq T \cdot N$; within these limits it makes no difference to cost or profit whether output is expanded in the t or in the n direction—for example, whether all machines are operated half time or half of the machines are utilized all available hours—as was obvious from the outset since the parametric production model gives x and the v's as functions not of t and n separately but of their product $t \cdot n$.[1] The shape of the cost curve will be the same as above.

---

[1] Still more generally, the time dimension may be split up into several components—for example, $t =$ number of days per month times number of hours per day—with a similar effect.

C. Cyclic Processes

(a) The models of the preceding sections, where the fixed factor is represented in the production function by its services in terms of hours per period, correspond to production processes which are *continuous* in the sense that output and inputs are continuous flows in time during effective operation. Let x units of output be produced in t hours, using the factor combination $v_1, \ldots, v_m$ ($v_1 = t$), and let τ denote "calendar" time during the t hours of effective operation ($0 \leq \tau \leq t$). Then output accumulates continuously in time according to a monotonically increasing function $\xi = F(\tau)$ (where $F(0) = 0$, $F(t) = x$), the instantaneous rate of output $d\xi/d\tau = F'(\tau)$ being positive for all τ.

When the production function is homogeneous of the first degree, $d\xi/d\tau$ is constant and equal to the average rate of output,

$$\frac{d\xi}{d\tau} = \frac{x}{t} = z, \quad \text{i.e.,} \quad \xi = F(\tau) = \int_0^\tau z\, d\tau = z \cdot \tau \tag{14}$$

where z is constant for a particular point of the production function. (If the inputs are substitutable, the same x can be produced by alternative combinations of t and z, one of which represents minimum cost, but this means changing the factor combination[1].) Output per hour, as well as inputs

---

[1] The fact that a continuous process is divisible in "calendar" time τ, as expressed by the continuity of the function $\xi = F(\tau)$, must not be confused with divisibility of the fixed factor's services in the time dimension represented by t, the number of hours of effective operation within a period of T hours. Calendar time is always divisible even when t is indivisible so that the plant has to be operated during the total time available ($v_1 = t = T = \bar{v}_1$, $0 \leq \tau \leq t$).

When v_1 is divisible and the capacity is not fully utilized in the time dimension ($t < T$), the t hours of effective operation need not be a continuous period; operation can be discontinued and resumed at will. "Calendar" time τ must be defined accordingly, being "suspended" during shutdowns.

The difference between the two time dimensions represented by t and τ can be illustrated in a factor diagram representing model (12a). Let P be the point of maximum profit. $\bar{x}$ units of output are produced in $t = T$ hours per period. Output and inputs accumulate in calendar time along the linear path $0\,P$, $\bar{x}/2$ units having been produced after $\tau = T/2$ hours of operation (point S), whereas $0\,B\,C$ (or $Q\,C$ if v_1 is indivisible) is the expansion path, R being the least-cost point if output *per period* were $\bar{x}/2$ (in which case the capacity would not be fully utilized in the time dimension, $t < T$).

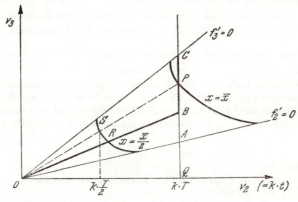

per hour, is constant during effective operation, each hour being an identical repetition of the preceding[1].

(b) Many important industrial processes, mechanical as well as chemical, are continuous in the above sense. Another important class of processes, of which numerous examples are to be found in the chemical industries, are the so-called *cyclic* processes. Whereas, in the case of continuous operations, the process is never shut down expect for inspection, repair, etc. (or because of dead time due to insufficient demand for the product), it is characteristic of cyclic operations that the process is discontinued periodically for discharging, cleanout, or reactivation, and the cycle of charging, operation, and shutdown is repeated for the number of cycles required to produce the desired quantity of product. The need for periodic shutdowns arises either when the instantaneous production rate decreases gradually with time so that the operation must be stopped and the initial conditions restored to give the original high production rate—in this case the cyclic process is referred to as semicontinuous—or when the nature of the process calls for production in batches, in which case no product is obtained until the unit is shut down for discharging[2].

Repetitive *batch* operations with fixed batch size and fixed cycle time present no particular problem: output accumulates in proportion to the number of cycles (batches) so that ξ is a step function of τ, the average overall rate ξ/τ being the same for all values of τ which are multiples of the constant cycle time.

In *semicontinuous* processes the time factor plays a more complicated part. Let w be the quantity of product made per cycle from a given amount of feed in operation time t_1. As shown in *Fig. 36*, w will be an increasing function of t_1,

$$w = \varphi(t_1), \tag{15}$$

but the instantaneous production rate $dw/dt_1 = \varphi'(t_1)$ is decreasing. The average production rate $w/(t_1 + t_2)$, where t_2 is cleanout time (constant), is a maximum—not necessarily representing an optimum—at the point (P) where it is equal to $\varphi'(t_1)$. [3]

For identical repetitions of the cycle, output will accumulate in calendar time (not counting dead or interim time) as shown in *Fig. 37*. Curve (I) represents $\xi = F(\tau)$ when the operation in each cycle is discontinued after six hours ($t_1 = 6$), to be followed by two hours of cleanout etc. ($t_2 = 2$).

---

[1] More generally, z may depend not only on some or all of the u_i but also on t:

$$z = \varphi(u_2, \ldots, u_m, t)$$

which, as we have seen, leads to a non-homogeneous production function. (For example, average output per hour may diminish with the length of the working day, t being the number of hours of operation per day.) In this case ξ will not be proportional to τ. Accumulated output after τ hours is $\xi = \tau \cdot \varphi(u_2, \ldots, u_m, \tau)$ and the instantaneous production rate is

$$\frac{d\xi}{d\tau} = \tau \cdot \frac{\delta\varphi}{\delta\tau} + \varphi$$

which is a function of τ.

[2] Cf., for example, SCHWEYER (1955), Ch. 9, esp. pp. 245 and 256.

[3] SCHWEYER, *op. cit.*, pp. 256 f.

Alternatively, if the process is shut down after three hours ($t_1 = 3$) and if $\varphi(3) = 0.75 \cdot \varphi(6)$, this will result in curve (II). The sooner the process is shut down in each cycle, the more cycles are required for a given quantity

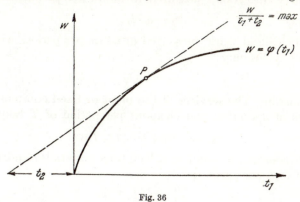

Fig. 36

of output per period of given length; the two curves lead to output $\bar{x}$ per day in three eight-hour cycles ($t_1 + t_2 = 8$, $n = 3$ where n is the number of cycles per period) or in four cycles of five hours' length.

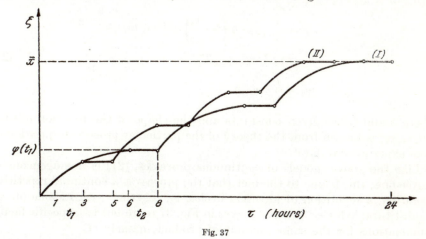

Fig. 37

(c) This kind of variation, which implies variation in the overall average rate of production

$$\frac{\bar{x}}{n \cdot (t_1 + t_2)} = \frac{\varphi(t_1)}{t_1 + t_2},$$

represents a factor substitution in disguise, as did variation of z in (14); the optimum factor combination for $x = \bar{x}$ depends on the way in which the relevant cost factors vary with the parameters t_1 and n. A parametric production model expressing output and inputs per period in terms of t_1 and n can be formulated as follows:

Each cycle produces $w = \varphi(t_1)$ units of output so that for n cycles per period we have

$$x = n \cdot \varphi(t_1). \tag{16}$$

Let v_1 represent the complex of variable inputs (except raw material) involved in active operation, measured in hours per period. Then we have

$$v_1 = n \cdot t_1, \tag{17}$$

whereas total cleanout work, measured in hours per period, will depend on the number of cycles only:

$$v_2 = n \cdot t_2 \tag{18}$$

where t_2 is constant. The services of the plant or fixed equipment, in terms of total hours of operation and cleanout per period of T hours, will be

$$v_3 = n \cdot (t_1 + t_2) \quad (\leq T). \tag{19}$$

Finally, for a constant amount k of feed (raw material) required per cycle, total input of feed per period is

$$v_4 = n \cdot k.^1 \tag{20}$$

Eliminating the parameters n and t_1 in order to translate (16)–(20) into a model involving economic variables only, we are left with a production function in (v_1, v_2) and two "shadow functions" expressing v_3 and v_4 in terms of v_1 and v_2:

$$x = \frac{1}{t_2} \cdot v_2 \cdot \varphi\left(t_2 \cdot \frac{v_1}{v_2}\right) = f(v_1, v_2)$$

$$v_3 = v_1 + v_2 \quad (\leq T) \tag{21}$$

$$v_4 = \frac{k}{t_2} \cdot v_2$$

where k and t_2 are given constants and the shape of the function $\varphi(t_1) = \varphi(t_2 v_1 / v_2)$ is known from the theory of the particular process in question or from experimental data[2].

Like the above models of continuous processes, (21) is homogeneous of degree one; this is due to the fact that the process is a continual repetitive operation, a repetition of identical cycles. v_2, together with its shadow v_4, is substitutable for v_1; the two curves in Fig. 37 represent two specific factor combinations for the same amount of product, namely (I)

$$\bar{x} = n \cdot \varphi(t_1) = 3 \cdot \varphi(6); \quad v_1 = n \cdot t_1 = 3 \cdot 6 = 18; \quad v_2 = n \cdot t_2 = 3 \cdot 2 = 6$$

and (II)

$$\bar{x} = 4 \cdot \varphi(3); \quad v_1 = 4 \cdot 3 = 12; \quad v_2 = 4 \cdot 2 = 8.$$

It is true that n can assume integral values only, but the resulting discontinuities in the variables are of little consequence when a sufficiently long period (T hours) is considered. The cyclic nature of the process then becomes

---

¹ Other inputs which are constant per cycle may be included in v_4 (or in v_2) which then becomes a complex factor.

² The parametric model (16)–(20) corresponds to the engineering model given by SCHWEYER, loc. cit., who does not, however, derive the economic production function (21). SCHWEYER also ignores (20) and the cost of material (feed).

a matter of secondary importance; v_1 and v_2 can be treated as continuously substitutable inputs and the curves $\xi = F(\tau)$ of Fig. 37 approach the linear shape (14) which characterizes continuous processes with a homogeneous production function.

This is brought out clearly if the process, like a continuous one, is analysed not on the basis of the cycle but in terms of output and inputs per *hour* of effective operation (or, to reduce the discontinuities, per day) and number of hours (days) per period. The parametric model then becomes

$$x = t \cdot z \qquad\qquad z = \frac{\varphi(t_1)}{t_1 + t_2}$$

$$v_1 = t \cdot u_1 \qquad\qquad u_1 = \frac{t_1}{t_1 + t_2}$$

$$v_2 = t \cdot u_2 \quad \text{where} \quad u_2 = \frac{t_2}{t_1 + t_2}$$

$$v_3 = t \cdot u_3 \qquad\qquad u_3 = 1$$

$$v_4 = t \cdot u_4 \qquad\qquad u_4 = \frac{k}{t_1 + t_2}$$

and where $t = n(t_1 + t_2) \leq T$.

(d) The least-cost factor combination for given $x = \bar{x}$ is determined by minimizing

$$c = q_1 v_1 + q_2 v_2 + q_4 v_4 = q_1 v_1 + \left(q_2 + q_4 \frac{k}{t_2}\right) v_2$$

subject to $x = f(v_1, v_2)$, treating v_2 and v_4 as a composite factor measured in units of v_2 (i.e., on an hourly basis). Because of the homogeneity of the model, the relative factor combination is independent of $\bar{x}$, the necessary condition for minimum cost being

$$(c' =) \quad \frac{q_1}{f_1'} = \frac{q_2 + q_4 k / t_2}{f_2'} \tag{22}$$

where $f_1' = \varphi'$ and $f_2' = (1/t_2) \cdot \varphi - (v_1/v_2) \cdot \varphi'$ are functions of v_1/v_2, cf. (21). Hence, as long as v_1 and v_2 are independent variables—i.e., within the capacity limit $v_1 + v_2 \leq T$—the expansion path is a straight line, $0A$ in *Fig. 38*, and marginal cost c' is constant[1]. Expansion along $0A$ represents a variation of capacity utilization in the time dimension: $t_1 = t_2 \cdot v_1/v_2$ is constant and so are z and the u_t so that output is expanded by varying the

---

[1] SCHWEYER (1955), p. 259, defines optimum operation as minimum *unit* cost, using t_1 as independent variable to find the minimum of

$$\frac{c}{x} = \frac{q_1 t_1 + q_2 t_2}{\varphi(t_1)}$$

(ignoring the cost of raw material). This is of course equivalent to minimizing total cost for given level of output, cf. (22), when marginal cost is constant for any given value of t_1.

total time of effective operation per period, $t = n(t_1 + t_2)$, that is, by varying the number of identical cycles, n.

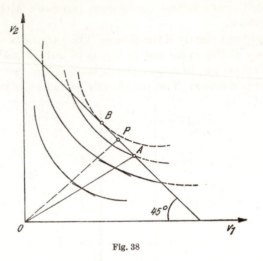

Fig. 38

At point A, capacity is exhausted in the dimension of t and further expansion will have to be along the capacity line $v_1 + v_2 = T$. For movements along the line, where $dv_1 = -dv_2$, we have

$$dx = (f_2' - f_1')dv_2 ;$$

hence, if the slope of the isocosts is greater than -1 (cf. Fig. 38), dx/dv_2 is positive at A and the expansion path will proceed from A to B, the latter point representing maximum output per period.

This kind of variation implies a change in the relative factor combination v_2/v_1; $v_2 (= n \cdot t_2)$ is increased at the expense of $v_1 (= n \cdot t_1)$, that is, output is expanded by increasing the number of cycles, but the length of the cycle $(t_1 + t_2)$ has to be reduced because the plant is already operated continually through the entire period T. [1] The time factor is involved in this kind of variation only in that less time is used for actual operation and more for cleanout, total time being constant.

Marginal cost along AB is

$$\frac{dc}{dx} = \frac{q_1 dv_1 + (q_2 + q_4 k / t_2) dv_2}{dx} = \frac{q_2 + q_4 k / t_2 - q_1}{f_2' - f_1'} \tag{23}$$

which at point A is equal to c' as determined by (22)[2]; as B is approached, dx tends to zero and dc/dx to infinity. Furthermore, marginal cost is monotonically increasing between A and B. Differentiating (23) with respect to x, recalling that $dv_1 = -dv_2$, we have

---

[1] Had the isocost lines been steeper than the capacity line AB, the relative positions of A and B in Fig. 38 would have been reversed: expansion along the capacity line would then require more of v_1 at the expense of v_2, i.e., fewer but longer cycles.

[2] If $a/b = c/d$, then $(a-c)/(b-d) = a/b$, as can be readily shown by cross multiplication.

$$\frac{d^2c}{dx^2} = \frac{-(q_2 + q_4 k / t_2 - q_1)}{(f_2' - f_1')^3} \cdot (f_{11}'' - 2 f_{12}'' + f_{22}'')$$

where

$$f_{11}'' = \varphi'' \cdot \frac{t_2}{v_2}, \quad f_{12}'' = \varphi'' \cdot \frac{-t_2 v_1}{(v_2)^2}, \quad f_{22}'' = \varphi'' \cdot \frac{t_2 (v_1)^2}{(v_2)^3},$$

i.e.,

$$f_{11}'' - 2 f_{12}'' + f_{22}'' = \varphi'' \cdot \left(\frac{t_2}{v_2} + \frac{2 t_2 v_1}{(v_2)^2} + \frac{t_2 (v_1)^2}{(v_2)^3} \right) < 0$$

so that $d^2c / dx^2 > 0$. Hence—neglecting the discontinuities in the variables due to n being an integer—the cost curve as defined along OAB has the same shape as that belonging to a homogeneous model of continuous production with a fixed factor which is divisible in the time dimension[1].

D. The Relevance of the U-Shaped Marginal Cost Curve

(a) The first conclusion to be drawn from all these examples is that, in a wide variety of apparently typical cases of production processes, the kind of cost function known from neoclassical theory is inapplicable because it does not represent the locus of least-cost combinations even when the production function has the shape typically assumed by the traditional theory of production. Neoclassical analysis, when based on the assumption that the fixed factor is indivisible in time and space, leaves out the important part which the extensive dimensions of capacity utilization play in the substitution process by opening up the possibility of proportionate variation of all inputs. The neoclassical cost function with the U-shaped marginal cost curve is relevant only in the special case where the plant is indivisible in space and its services are also indivisible in the time dimension.

It has long been recognized that the rate of capacity utilization may have several dimensions, that in such cases the kind of variation considered must be specified in order for the cost function to be defined—this aspect is particularly important in the interpretation of empirical cost curves—and that the optimum path of expansion and the resulting cost function will involve variation in all dimensions[2]. None the less, the U-shaped marginal

---

[1] Again, output and inputs accumulate in *calendar time* along the line OP in Fig. 38 where P represents the factor combination actually chosen (the point of maximum profit); cf. above.

[2] WINDING PEDERSEN (1933), in an attempt to reconcile the neoclassical cost curve with the linear cost function assumed by the "engineering" school of thought (notably JANTZEN (1924)), was apparently the first to draw attention to the significance of divisibility in the time dimension, interpreting the two types of cost functions as resulting from variations in the intensive and the time dimension of capacity utilization respectively. This idea was further elaborated by SCHNEIDER (1937, 1940), which led v. STACKELBERG (1941), esp. p. 36, to the conclusion that the cost curve, as defined by cost minimization for parametric level of output, will have the general shape of Fig. 32 above. The same discovery was made independently by DEAN (1941), pp. 5 ff., who also pointed out that physical divisibility of the plant will have a similar effect on the shape of the cost curve. (DEAN (1951), p. 275, has suggested a third source of cost linearity, namely variability of machine speed; however, as model (10)–

cost curve has shown great persistence in economic literature. Several explanations of this fact may be offered.

For one thing, neoclassical theory has been traditionally preoccupied with production processes which are not continuous in the above sense. In the absence of empirical material on industrial production processes, factor substitution has been illustrated largely by examples drawn from agriculture, the classic example being grain cultivation with land as the fixed factor. In processes of this type the product is obtained intermittently, the instantaneous rate of output being zero until the crop is ripe and ready to be harvested; the production period, therefore, is indivisible and the utilization of the fixed factor has a time dimension only in the trivial sense that the process is capable of repetition during the next period. (It is precisely for this reason that, in such cases, the dimension of output and inputs is often indicated in terms of absolute physical units, neglecting "per period.") During the production period the capacity of the "plant" is always fully utilized in the time dimension. If the spatial divisibility of the fixed factor is also disposed of by defining the production function as referring to a given plot of land all of which is in cultivation, the shape of the production function and that of the cost curve are likely to follow the law of variable proportions.

Secondly, it has not always been realized that, even when a factor of production is an indivisible unit, its services may be perfectly divisible. A machine as such is an indivisible fixed factor, but when the rate of capacity utilization can be varied in the time dimension, the relevant input in the production function is not the number of machines but their services in terms of number of machine hours used per period. The input of machine services is continuously variable up to capacity so that it is a scarce input rather than a fixed factor. The distinction between the factor and its services is less important when the use of the factor is indivisible in the time dimension; number of acres or acre services per year will do equally well. But in any case of divisibility—spatial or temporal, or both—the capacity restriction must have the form $v_1 \leq \bar{v}_1$ and the amount actually used, v_1, must be specified in the production function.

One reason why this point has often been missed is, perhaps, that v_1 usually does not appear in the cost function because no positive price is associated with this input. Another is that inequalities tend to be mathematically awkward in marginal analysis. It is true that the introduction of a single capacity restriction in inequality form—such as the capacity limit in model (12a)—does not unduly complicate the argument, but the resulting

---

(11a) above shows, higher machine speed requires more energy or fuel per machine hour so that this is really a case of more intensive operation with changing factor proportions, cf. WINDING PEDERSEN (1949), p. 51.)

The optimal expansion path was indicated in a very general way by SCHNEIDER (1937) as the locus of tangency between "isocosts" and "isoquants" in a (z, t) diagram. The specific shape of the path—i.e., constant z up to the point where $t = T$—and the corresponding cost curve (Fig. 32) were subsequently developed by SCHNEIDER (1942), pp. 148 ff., and illustrated by an example—based on engineering data—of the type (10)–(11a), cf. SCHNEIDER (1958), pp. 170 and 189 ff. (see Ch. VII, B above).

cost function (Fig. 32) is not an analytic function; it is pieced together from two such functions, in contrast to the cost function derived from the law of variable proportions. It was only the discovery of the linear models of production which paved the way for inequality constraints in the theory of production: in order to get meaningful solutions to linear optimization problems, methods had to be devised for maximization of linear functions subject to linear inequalities[1].

(b) Needless to say, the question of the practical relevance of the U-shaped marginal cost curve cannot be settled by theoretical argument alone: it is ultimately a matter of empirical research. It must be emphasized, however, that the underlying question of factor divisibility is not a purely technological one. While some cases of indivisibility in the time dimension are due to technical necessity (processes which cannot be discontinued during the period), there are other cases where shutdowns are technically possible but undesirable for different reasons, for example, as a matter of policy[2]. As regards spatial divisibility, agricultural cost functions will have a linear segment if the acreage in cultivation is adjusted optimally to the level of output, but the marginal cost curve is likely to be U-shaped, despite the physical divisibility of the factor "land," if it refers to a given plot in cultivation[3]. In short, the result will depend on the underlying assumptions with respect to the fixed factor, and the same technology may lead to either type of cost function.

---

[1] Cf. Appendix 1.

[2] Cf. *Cost Behavior and Price Policy* (1943), p. 111 n.

[3] This is the situation that underlies empirical production and cost functions derived from experiments in agriculture.

Chapter VIII

Product Quality and the Production Function

A. Product Quality in the Theory of Production

We have so far disregarded the complications that my arise if variations in product quality are taken into consideration.

The concept of a production function has operational meaning only with reference to a particular product or class of products; in order for the model to be defined at all the product must be specified in terms of its relevant properties. When nothing explicit is said about product quality and the quantity produced is considered the sole dimension of the product, the specification must be understood to be implicit in the model, it being tacitly assumed that the general nature of the product as well as its specific quality is given by the selection of inputs and the shape of the production function. To the extent that the qualitative properties of the product reflect the properties of the inputs, a sufficiently uniform product quality can be ensured by excluding such inputs from consideration as would lead to an unacceptable product (e.g. inferior raw materials)[1]; and the production function may be defined as the set of points which represent a satisfactory quality as specified by the producer. Different grades of the "same" product may of course exist but are treated as so many distinct products, each of which has a separate production and cost function (unless they are produced jointly, a case to be described in terms of a multi-product model).

The development of the theory of the firm that took place in the thirties led to the recognition that product quality—as well as selling effort—is an important parameter of action to the firm along with price (or quantity produced)[2]. The emphasis, however, was on the demand side: it was of course realized that differences in product quality are reflected in the costs of production, but little attention was paid to the underlying production function and to the way in which the cost function is derived from it when

---

[1] It will be recalled that our definition of the production function (Ch. II) requires each particular input to be qualitatively well-defined; qualitative change in one or more of the specified factors represents technological change, i.e., a different production function. This does not rule out the possibility of two or more inputs in the production model representing different grades of the "same" input (e.g. high-quality and low-quality materials which may be substituted for one another without seriously affecting product quality), only each of them must be represented by a separate variable v_i.

[2] The pioneer work was CHAMBERLIN (1933).

product quality occurs as a parameter. This was natural enough as long as product quality was regarded as non-quantitative[1]: in that case the proper way of introducing the quality aspect is to consider a multitude of different production and cost functions, each of which corresponds to a particular product quality—in effect, to a different product.

The subsequent recognition that product quality is a multi-dimensional concept defined by a number—possibly a large number—of quality criteria, some of which can and should be regarded as quantitative, measurable magnitudes[2], made no appreciable change in the picture so far as non-quantitative criteria were concerned: each particular *combination* of such criteria should obviously be considered as a separate product with a production and cost function (as well as a demand function) of its own[3]. It is only when the product cannot be exhaustively described in terms of these criteria that a problem arises. Clearly the quantitative criteria must somehow appear as parameters in the cost function—higher quality is likely to be associated with higher cost—but the relationship of the quality parameters to the underlying production function is by no means unambiguous.

We shall now subject the quality problem to a more rigorous treatment, in order to examine the various ways in which quality parameters may come into the production model and the effects of quality specifications on the problem of input substitution and cost minimization. Since the analysis is largely concentrated on the production and cost relationships, no attempt will be made to explain how the optimum level of product quality is actually determined; this is a problem to be solved within a wider frame of reference, that of the theory of the firm in its most general sense, which provides a model for the simultaneous determination of output, price, product quality, selling effort, and inputs[4]. In the present context, product quality will be treated as a parameter, or a set of parameters, to be specified somehow by the firm.

By product quality, then, we will understand the total of such physical and other characteristics of the product as affect the cost of producing it; in other words, quality is the set of properties needed for a complete specification of the product and for the cost function to be determined[5]. Only quantitative, measurable quality criteria will be considered; properties which are non-quantitative in the sense that they cannot be represented by a continuously variable parameter will be disregarded (i.e., assumed to be

---

[1] CHAMBERLIN, *op. cit.*, pp. 78 f., then held the view that "'product' variations are in their essence qualitative rather than quantitative" and therefore non-measurable.

[2] BREMS (1951) is the chief exponent of this school of thought.

[3] Cf. BREMS, *op. cit.*, p. 57.

[4] The reader is referred to the special literature, in particular to BREMS (1951). Some useful references are given in BREMS (1957), p. 105.

[5] From the point of view of demand, on the other hand, the product and its qualities are defined by specifying the properties which are represented by parameters in the demand function. The two sets of quality parameters do not necessarily coincide and a complete model of the firm should take all quality dimensions into account, whether they appear in the demand or cost function or in both.

implicit in the shape of the production function) since, as we have seen, products which differ in respect to such qualities may in the present context be treated as distinct commodities.

B. Quality Parameters in the Production Function

(a) In order for the input-output relationships to be fully specified, the product must be completely defined in terms of quality: all relevant quality dimensions must be specified, whether they are measurable or not.

Some kinds of quality change—particularly those concerning non-quantitative quality criteria—can be effected only by discontinuous change in the technology, that is, by switching to a different production function. On the other hand, as to such dimensions as are quantifiable, it seems plausible to assume that a continuous range of quality levels—as represented by the values of the continuous quality parameters—can be produced within the same basic technology[1]; the same inputs are used but higher product quality, like a higher rate of output, requires more of some or all inputs.

This means that the quality parameters appear as parameters in the production model; or, to put if differently, the technology of the process can be described by a family of production functions. In practical terms, the interpretation is that the shape of the production function represents an incomplete description of the technical procedure by which a given selection of well-defined inputs are transformed into the product; the "details" of the technology are assumed to depend uniquely on the particular product quality desired. Once the values of the quality parameters have been fixed by the firm, the production relationships are fully specified; conversely, product quality is continuously manageable—at least within limits—through modifications in the technical procedure adopted by the firm.

By way of illustration, let output x depend on a single input v and a quality parameter y, assumed to be continuously variable:

$$x = x(v, y).$$

The amount of input required depends not only on the quantity to be produced but also on the quality dimension of the product as represented by y. Within the relevant region, x is an increasing function of v, i.e.,

$$\frac{\delta x}{\delta v} > 0 \qquad (y \text{ constant}),$$

and so is y,

$$\frac{\delta y}{\delta v} = - \frac{\delta x / \delta v}{\delta x / \delta y} > 0 \qquad (x \text{ constant}),$$

assuming that it takes more of the input to produce a better quality (as represented by a higher value of y)[2]. For given v, output can be increased

---

[1] ZEUTHEN (1928), p. 30.

[2] ZEUTHEN, loc. cit., used the term "marginal quality improvement" for $\delta y / \delta v$, though he had a somewhat different production model in mind (cf. below).

only at the expense of quality and vice versa, $\delta x / \delta y$ (v constant) being negative. Since v is a function of the quantity produced as well as of the quality dimension of the product, the quality parameter also appears in the cost function:

$$c = q v = c(x, y)$$

(where $\delta c / \delta x$ and $\delta c / \delta y$ are > 0).

(b) More generally, let us assume that the technology of a process can be described by a family of production functions with H independent quality parameters $y_1, y_2, \ldots, y_H$ and m substitutable inputs:

$$x = x(v_1, v_2, \ldots, v_m; \ y_1, y_2, \ldots, y_H). \tag{1}$$

Once a set of numerical values have been assigned to the quality parameters,

$$y_h = \bar{y}_h \qquad (h = 1, 2, \ldots, H), \tag{2}$$

the commodity is specified and a production function of the neoclassical type is established.

In the case of only two inputs and one quality parameter, y, the factor combinations which will produce a given quantity of output $\bar{x}$ can be illustrated geometrically by a family of isoquants, each one corresponding to a particular level of quality, as shown in the following diagram (*Fig. 39*), where $\bar{y}$ and $\bar{\bar{y}}$ ($\bar{\bar{y}} > \bar{y}$) represent two different values of y.

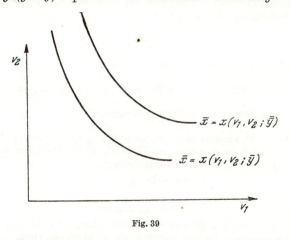

Fig. 39

When no explicit attention is paid to the quality aspect, the specification of the quality parameters that is needed in order to fix the position of the isoquants, and thus also to determine the least-cost points and the cost function, must be understood to be given before the production function $x = x(v_1, \ldots, v_m)$ is written down; it is implicit in the shape of the function. With a view to quality variation, however, the problem of cost minimization can be solved with the y_h as provisionally unspecified parameters along with x. Minimizing total cost

$$c = \sum_{i=1}^{m} q_i v_i$$

subject to (1) for given x and given y_h (and for given factor prices), the necessary conditions are

$$(u=)\ \frac{q_1}{x_1'} = \cdots = \frac{q_m}{x_m'},$$

which together with (1) determine the v_i in terms of the given x and y_h, that is, for all conceivable values of x and the y_h. Substituting the solution to this parametric minimization problem in the cost expression we have

$$c = \sum_{i=1}^{m} q_i v_i = c(x; y_1, y_2, \ldots, y_H). \qquad (3)$$

This is the parametric cost function—or family of cost functions—usually considered in price-quality equilibrium models of the firm[1]. For fixed values of the y_h, (3) becomes a cost function of the familiar type.

(c) Quality criteria may of course appear as parameters in other types of production models, for example, in a model characterized by fixed coefficients of production[2]:

$$v_i = a_i x \qquad (i = 1, 2, \ldots, m).$$

For given product quality the coefficients a_i are constants, but different quality specifications will correspond to different values of the a_i, which are therefore functions of the y_h[3]:

$$a_i = a_i(y_1, y_2, \ldots, y_H);$$

hence the production model can be written as

$$v_i = a_i(y_1, y_2, \ldots, y_H) \cdot x \qquad (i = 1, 2, \ldots, m), \qquad (4)$$

and the corresponding cost function becomes

----

[1] BREMS (1951), p. 56, explains the occurrence of quality (and selling-effort) parameters in the cost function in this manner, and proceeds to show (pp. 62 ff.) how the optimal values of x, the y_h, and p are determined by maximizing profit

$$z = px - c(x; y_1, \ldots, y_H)$$

subject to the demand function

$$x = f(p; y_1, \ldots, y_H)$$

(where the product price p is a parameter of action along with the y_h).

[2] This has been suggested by BREMS (1957). Ch. 17 of BREMS (1959) contains a revised version of this article.

ZEUTHEN (1928), p. 30, pointed out that, in some cases of fixed coefficients of production, partial variation in one input may affect product quality though the level of output remains the same. This will be so in a model of the type

$$v_1 = a_1(y) \cdot x, \qquad v_2 = a_2 \cdot x$$

where, for v_2 and thus also x constant, changes in v_1 will affect y. (Had both coefficients of production been functions of y, as in model (4) below, v_2 and x could not both have been constant for variations in v_1.)

[3] For example, the amount of labour required per unit of product may depend on the desired degree of workmanship, cf. BREMS (1957), p. 106. This quality criterion may of course also appear in production functions of the type (1), where labour is substitutable for other inputs.

$$c = \sum_{i=1}^{m} q_i a_i x = \bar{c}\,(y_1, y_2, \ldots, y_H) \cdot x.^1 \tag{5}$$

Since the a_i are uniquely determined when product quality has been specified by assigning a set of particular values to the y_h, the coefficients of production may themselves be taken as representing product quality, at least in the sense that a change in a_i reflects a change in the quality of the product; quality is then defined as the complete specification of the production process[2]. This approach has the advantage of evading the somewhat tricky problem of quality measurability in a simple and elegant manner; the a_i are always measurable.

(d) In deriving the cost functions (3) and (5) we have treated the quality dimensions y_h as provisionally unspecified parameters, to be determined in a comprehensive model (not dealt with here) which takes into account the effect of quality variation on the demand for the product[3]. If the firm considers this effect to be negligible—this may be the case if demand is largely dependent on the non-quantitative quality criteria which are implicitly specified in the shape of the production function—the firm will profit by fixing the y_h at their lowest possible values[4]. Or the market may be such that the product is judged to be acceptable only if its quality is kept within certain limits,

$$\bar{y}_h \leq y_h \leq \bar{\bar{y}}_h,^5 \tag{6}$$

in which case the cost of production is at a minimum for any given x if each of the quality parameters is fixed at its lowest permissible limit, $\bar{y}_h$.

C. Quality Constraints

1. Quality Functions and the Production Function

(a) The purpose of the models above has been to show that the appearance of quality parameters in the production function may be due to a technology with "built-in" possibilities of quality variation. The quality dimensions of

---

[1] It might be argued that (4) is not a case of fixed coefficients (limitational inputs) since the a_i can in fact be varied. The point is, however, that such variations cannot be interpreted as input substitution (accompanied by quality change) since they are brought about not by a change in factor prices but by variations in product quality.

[2] Cf. BREMS (1957), p. 106; this observation is the main point in the article.—Strictly speaking, it will have to be assumed that each a_i depends on one quality parameter only, since otherwise there may be different combinations of the y_h which correspond to the same value of a_i (i.e., the quality parameters may be "substitutes" with respect to the coefficient of production) so that product quality may change without a_i being affected. However, it is scarcely to be expected that such a change in the y_h will leave all of the a_i unaffected.

[3] Cf. above.

[4] Assuming that the v_i for given x—as given by the technology or determined by cost minimization—and thus also the cost function are monotonically increasing (or at least non-decreasing) functions of the y_h.

[5] This way of taking the demand function into account amounts to assuming that demand will respond in a discontinuous manner to quality changes that go beyond certain limits.—Alternatively, (6) may have a purely technical interpretation: the same basic process permits of quality variation only within limits.

the product enter as parameters in the production function (more generally: in each of the relations that constitute the production model) so that the product has to be completely specified before the technological model can be established.

However, this is not the only possible explanation of a cost function depending on quality dimensions as well as on output. It is conceivable that, for a continuous range of product variants, the production function as such is independent of the quality dimensions—in order to specify the technology it is sufficient to indicate roughly the general nature of the product—but that the quality of the product varies with the factor combination[1]. In other words, in addition to the production function proper, e.g.

$$x = x(v_1, v_2, \ldots, v_m), \tag{7}$$

the complete production model includes a number of additional equations expressing the quality dimensions in terms of the inputs:

$$y_h = y_h(v_1, v_2, \ldots, v_m) \qquad (h = 1, 2, \ldots, H);[2] \tag{8}$$

for given specification of the product,

$$y_h = \bar{y}_h, \tag{9}$$

equations (8) impose constraints on the variations of the v_i. [3] From a formal point of view (7)–(8) may be regarded along with the previous models as a multi-equation production model containing quality parameters. It is a special case, however, in that the technology of the process, as represented by (7), is separated from the quality aspect; product quality is controlled only through the choice of factor combination.

Two particular cases of this model can be distinguished according as to whether or not (8) contain inputs which do not occur in (7)[4].

(b) Let us first assume that there are no other variables (inputs) in (8) than those already present in (7)[5]. In that case (8) can be described as pure quality constraints: for specified values of the y_h they are side relations which rule out certain areas of the production function[6] (7). For a sufficiently

---

[1] Cf. BARFOD (1936). This important article, which was the first serious attempt to treat the problem of quality variation within the context of the theory of production, appears to have passed almost unnoticed.

The same general idea has later been suggested by CHENERY (1953), p. 310, possibly inspired by linear-programming models of industrial blending processes where quality specifications lead to side conditions on the inputs (cf. below).

[2] There is no need to include x among the variables of (8) since it can always be eliminated, using (7).

[3] In BARFOD's terminology, (7) is called "the quantity function" and (8) "the technical quality function", *op. cit.*, pp. 27—29.

[4] These cases—dealt with under (b) and (c) below—correspond to BARFOD's cases I and III, cf. *op. cit.*, p. 42. BARFOD's model II is a special case of III, cf. below.

[5] This assumption does *not* imply that *all* of the v_i in (7) also occur in (8). It is perfectly possible for product quality to depend on some of the inputs only (e.g. those representing materials).

[6] CHENERY, *loc. cit.*

large number of quality specifications (9)—i.e., for $H = m - 1$—substitution is ruled out entirely, the model being deprived of the degrees of freedom required for economic choice.

By way of illustration, let us assume that there are only two inputs and one quality criterion, y. For given level of output $\bar{x}$, (7) defines an isoquant representing the *technical* range of substitution as shown in *Fig. 40*.

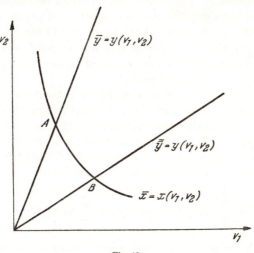

Fig. 40

With each point on the isoquant is associated a particular level of quality, given by (8). If y (instead of x) is given, (8) defines an iso-quality curve in factor space.

If product quality is rigidly specified, $y = \bar{y}$, the range of substitution is in the two-factor case reduced to a point A which represents the intersection of the isoquant and the iso-quality curve, so that v_1 and v_2 become limitational inputs. A less rigid quality specification in terms of a pair of upper and lower limits, $\bar{y} \leq y \leq \bar{\bar{y}}$, will permit of some variation in the relative factor proportion as shown in Fig. 40, where points between A and B represent factor combinations which are compatible with the quality requirements as well as with the technical production function[1]. The least-cost point, as determined by minimizing cost subject to $x = x(v_1, v_2)$ and $\bar{y} \leq y(v_1, v_2) \leq \bar{\bar{y}}$ for given factor prices, will be either a boundary point (A or B, where $y = \bar{y}$ or $y = \bar{\bar{y}}$ respectively) or a point somewhere between A and B, that is, the point representing minimum cost when the quality limits

---

[1] The imposition of an *upper* limit to product quality is not as unrealistic as it might appear, if it is recalled that a higher value of y_h cannot always be interpreted as "better" quality (in the sense that it is valued higher by consumers); some of the y_h may represent "technical" quality dimensions which should be kept within certain limits (tolerances) in order for the product to be satisfactory. The product may also be specified in this way merely in order to distinguish it from other products. In fact, specifications of this kind—to be interpreted as concessions to consumers' preferences, a simplified way of taking the dependence of demand on product quality into account—are no more remarkable than exact specifications, where the two limits coincide.

are disregarded. Such a tangency minimum is the more likely, the wider the quality tolerance is and the less product quality depends on the relative factor combination.

In the general m-factor case, the introduction of H quality specifications will reduce the number of independent input variables from m to $m - H$ and the complete production model (7)–(9) can be described as a case of constrained substitution[1]. For given x the range of substitution in m-dimensional factor space is reduced to the configuration of points which represents the intersection of the isoquant—as defined by (7) for $x = \bar{x}$—and the isoquality surfaces (8)–(9)[2]. If product quality is specified only in terms of pairs of inequalities, the permissible region of substitution is represented by a bounded area on the isoquant surface.

If the quality dimensions are not specified beforehand but are regarded as unknown parameters of action whose optimal values are to be determined in a wider model which takes the demand function into account, the least-cost point for given x and the cost function can be determined with the y_h as parameters. Minimizing total cost subject to (7)–(8) for given $x = \bar{x}$ and $y_h = \bar{y}_h$, differentiation of the Lagrangian expression

$$L = \sum_{i=1}^{m} q_i v_i + u \cdot (\bar{x} - x) + \sum_{h=1}^{H} u_h \cdot (\bar{y}_h - y_h)$$

with respect to the v_i leads to m necessary conditions which with (7) and (8) determine the v_i (and the Lagrangian multipliers u and u_k) in terms of $\bar{x}$ and $\bar{y}, \ldots, \bar{y}_H$. Substituting in $c = \Sigma q_i v_i$ we get a parametric cost function

$$c = c(x; y_1, y_2, \ldots, y_H) \tag{10}$$

---

[1] Cf. Ch. V, B.

[2] BARFOD, *op. cit.*, pp. 49—51, cites an empirical example from the technical literature (cf. FOG (1934), pp. 33 ff.). The process in question is the decolorization of vegetable oils by means of bleaching coal or bleaching earth, or by a mixture of both. These bleaching agents remove the colour constituents from the oil by adsorption. Let v_1 and v_2 denote the quanties (in kg.) of the two bleaching agents applied to v_3 kg. of unbleached oil, and let x be the output (in kg.) of bleached oil. The production function for the bleaching process consists of two equations. The first is a material balance

$$x = v_3 - T_1 v_1 - T_2 v_2$$

where T_1 and T_2 are constants—determined by experiment—expressing the amounts of raw material (unbleached oil) lost in the adsorption process per kg. of bleaching agent applied. (BARFOD erroneously writes the material balance as $x = v_3$, but does take the losses into account when solving the optimization problem.) Second, the degree of bleaching obtained—as expressed by the extinction coefficient, y, which can be measured by means of a photometer—is a function of the three inputs,

$$y = y(v_1, v_2, v_3),$$

which puts an additional constraint on the variables when the quality of the product—the desired degree of bleaching—is specified, $y = \bar{y}$. For a given amount of v_3 to be bleached, a given value of y can be obtained by alternative combinations of v_1 and v_2, which are therefore substitutes in producing this quality effect. The shape of the function y for given v_3 under given process conditions has been determined experimentally for cocoanut oil and soybean oil respectively and illustrated by "isocolors", which permit the graphical determination of the optimal combination of the bleaching agents for given prices.

just as we did in the case of production functions with "built-in" quality parameters (cf. (3) above).

On the other hand, the firm may choose to disregard the effect of quality variations on the demand for the product and allow quality to be determined solely by cost considerations. In that case the production model consists of (7) only and the cost function is determined by minimizing cost subject to (7) for given parametric x, taking full advantage of the unconstrained range of substitution along the isoquants. The values of the y_h, as determined by inserting the least-cost factor combinations in (8), represent the cheapest quality level for given x.

A similar procedure may be resorted to if demand conditions are taken into account only by specifying a set of upper and lower limits to product quality, cf. (6) above. If the values of the y_h determined by minimizing cost for given x (or, more generally, by maximizing profit) subject to (7) turn out to respect inequalities (6), there is nothing to add; on the other hand, if one or more of the inequalities is violated, the problem has to be reformulated as a case of constrained maximization with (6) and (8) as boundary conditions and the optimal solution will be a point on the boundary of the permissible region.

Quality constraints of the type (8) may be combined with other types of technological models than (7), in which all inputs were assumed to be continuously substitutable. Substitution may be discontinuous, or some of the inputs may be limitational. However, some degree of substitution must be present in the purely technical production relations if quality variation is to be possible[1].

(c) The second class of cases is characterized by quality functions (8) containing inputs which do not appear in the production function (7) or, more generally, in the technological relations which constitute the production model proper. In cases of this type equations (8)–(9) can no longer be described as pure quality constraints; they also serve to introduce additional inputs[2] and to link them up with those of (7), i.e., they contain technical information and in this sense are part of the technological production model. Thus a complete separation of technology and quality considerations is no longer possible as it was in the previous class of models; in fact, (7)–(8) may be thought of as a multi-equation production model with built-in quality parameters similar to the cases dealt with above. The model represents a special case only in that quality parameters do not occur in all the relations of the model.

By way of example, consider the model

$$x = x(v_1, v_2, v_3) \tag{11}$$

$$y = y(v_1, v_2, v_3, v_4, v_5). \tag{12}$$

---

[1] In a model where all inputs are limitational, quality variation is possible only if the coefficients depend on quality parameters, but this means that we have a model where the y_h are built into the production relationships, cf. (4) above.

[2] In BARFOD's terminology such inputs are called "technical quality factors"; cf. BARFOD, op. cit., p. 42.

Since the production function (11) does not contain the quality parameter y, it is possible to produce the commodity using only v_1, v_2, and v_3 if product quality is disregarded; the cost function is then determined by minimizing total cost of the three inputs subject to (11) for parametric $x = \bar{x}$, and the quality level of the resulting crude product is given by (12) for $v_4 = v_5 = 0$. If a better product is desired, quality can be improved by use of v_4 and v_5 (assuming y_4' and $y_5' > 0$); for given specified $y = \bar{y}$, equation (12) becomes part of the production model, a kind of generalized "shadow function" with v_4 and v_5 as factor shadows[1]. Unlike the quality constraints dealt with above, (12) does not reduce the range of substitution defined by (11) for given x, but the position of the least-cost point will depend on the specified value of y. Minimizing the cost of all five inputs subject to (11)–(12) for parametric x and y we get the necessary conditions

$$q_i - u_1 \cdot x_i' - u_2 \cdot y_i' = 0 \qquad (i = 1,\, 2,\, 3)$$
$$q_i \qquad\qquad - u_2 \cdot y_i' = 0 \qquad (i = 4,\, 5)$$

(u_1 and u_2 being Lagrange multipliers) which with (11)–(12) determine each of the v_i, and thus also cost, as functions of x and y.

Had product quality been a function of v_4 and v_5 only,

$$y = y(v_4, v_5), \tag{12a}$$

the optimal combination of v_1, v_2, and v_3 would have been independent of y, their values being determined by (11) for given $x = \bar{x}$ and the conditions

$$q_i - u_1 \cdot x_i' = 0 \qquad (i = 1,\, 2,\, 3),$$

whereas v_4 and v_5 would depend solely on the specified quality level, being determined by

$$q_i - u_2 \cdot y_i' = 0 \qquad (i = 4,\, 5)$$

together with (12a) for $y = \bar{y}$.

This special case, however, is not likely to occur since, if the quality constraints have no variables in common with the technological production function, there is no connexion in the model between the quality factors and the quantity of output produced. In general, a higher level of output must be expected to require greater amounts of these inputs[2]. In model (11)–(12) this link is provided by the occurrence of v_1, v_2, and v_3 in both functions. A more direct connexion is established if output x takes their place in the quality function,

$$y = y(x, v_4, v_5) \tag{12b}$$

(which can be written in the form (12), substituting (11) for x)[3]. In cases of this type, like (12a), the optimal combination of v_1, v_2, and v_3 for given x is independent of the specified value of y, but the optimal values of the quality factors are functions of x as well as of y.

---

[1] Cf. Ch. V, A.

[2] BARFOD, *op. cit.*, who treats models of this type as a separate case (II), appears to have overlooked this point.

[3] For given y, model (11)–(12b) becomes an example of what FRISCH has called "disconnected factor rings," cf. Ch. V, C above.

Quality constraints of the type (12), with the special cases (12a) and (12b), are characterized by containing several quality factors—in the present models, v_4 and v_5—which are substitutable for each other with respect to product quality. There may well be cases, however, in which each particular property of the product is provided by one particular quality factor only so that in addition to (11) we have quality constraints of the form

$$y_1 = y_1(x, v_4)$$
$$y_2 = y_2(x, v_5)$$
(12c)

where v_4 and v_5 become limitational shadow factors (product shadows)[1] when the quality of the product has been specified[2][3]. In cases of this type the position of the least-cost point is independent of the quality specifications, depending solely on the prices of the inputs in (11).

2. Quality Constraints in Blending Processes

(a) Quality functions of the type (8) are of particular importance in blending processes, where the product is made by mixing a number of materials[4]. The general nature of the product is determined by the choice of materials, but the particular quality of the blend will usually vary in a continuous manner with its composition; indeed, the very purpose of blending ingredients is to get a product which in a sense represents a combination of their respective properties.

The basic relation of a production model representing a physical blending process is a simple material balance,

$$x = \sum_{i=1}^{m} v_i,$$
(13)

where the v_i are the quantities of the respective ingredients (materials)[5]. This equation, which is a special form of the production function (7), hardly conveys any technical information except the trivial fact that the materials are equivalent (perfect substitutes) as far as the quantity of output is con-

---

[1] Cf. Ch. V, A above.

[2] One interpretation of (11)–(12c) is that (11) represents a preliminary operation producing a crude product which subsequently undergoes further treatment (finishing operations) requiring specific inputs to provide the qualities desired in the finished product.

[3] In the case of fixed coefficients of production, (12c) can be written in the form

$$v_4 = a_4(y_1) \cdot x, \qquad v_5 = a_5(y_2) \cdot x.$$

The coefficients a_4 and a_5, which are uniquely determined by the quality parameters, may take the place of y_1 and y_2 as representing product quality, cf. above.

[4] The mixing of ingredients with the purpose of bringing about a chemical reaction is not treated here; we are concerned only with physical blending operations where the components preserve their chemical identity.

[5] (13) always holds if x and the v_i are measured in units of weight, whereas volumes are not necessarily additive; for example, a change of volume occurs if water is mixed with alcohol.

cerned. Clearly this is not a very interesting case—the least-cost factor combination consists in using the cheapest ingredient only—unless further restrictions are imposed by quality specifications.

Let $y_1, y_2, \ldots, y_H$ denote measurable quality criteria, and assume that each quality dimension of the product depends on the blending proportions, i.e., on the v_i/x:

$$y_h = y_h\left(\frac{v_1}{x}, \frac{v_2}{x}, \ldots, \frac{v_m}{x}\right) \qquad (h = 1, 2, \ldots, H), \qquad (14)$$

which may be written in the form (8) if the material balance (13) is substituted for x. When the desired quality of the product is specified either by assigning particular values to the functions or by specifying upper and lower limits to the y_h, (14) impose constraints—equations or inequalities—on the choice of factor combination.

The most restricted case is that of $H = m - 1$ quality specifications in equality form, where the v_i are uniquely determined by (13)–(14) for given x. This may happen, for example, if each ingredient is the sole provider of a particular quality dimension so that the constraints have the form

$$y_h = y_h\left(\frac{v_h}{x}\right) \qquad (h = 1, 2, \ldots, H) \qquad (14a)$$

which for $y_h = \bar{y}_h$ $(h = 1, 2, \ldots, m - 1)$ determine the blending proportions[1] so that the production function can be written

$$v_h = a_h \cdot x \qquad (h = 1, 2, \ldots, m)$$

where

$$a_h = y_h^{-1}(\bar{y}_h) \qquad (h = 1, 2, \ldots, m - 1)$$

and, by (13),

$$a_m = 1 - \sum_{h=1}^{m-1} a_h.$$

In all other cases—for fewer than $m - 1$ exact quality specifications or for any number of quality requirements in inequality form—a range of substitution exists and the optimal composition of the blend is determined by minimizing cost subject to (13) for given output and to (14) for specified values or permissible intervals of the y_h.

The quality functions will frequently be of linear form,

$$y_h = \sum_{i=1}^{m} y_{hi} \cdot \frac{v_i}{x} \qquad (14b)$$

where the y_{hi} are constants[2] to be interpreted as quality indicators for the respective inputs[3] so that, for each of the specified quality dimensions,

---

[1] Had all m quality dimensions been specified, the model would have been overdetermined, cf. (13).

[2] Sometimes called "blending numbers," cf. Symonds (1955), p. 4.

[3] For example, if the product is made of the first material only so that $v_1 = x$, (14b) identifies y_{h1} with y_h.

product quality y_h is a weighted average of the qualities of the ingredients with the blending proportions as weights[1]. This will always be so when y_{hi} represents the *content* of some constituent in a unit of v_i: the total content of the constituent per unit of x, i.e., in a unit of the blended product, is indicated by the right-hand side of (14b)[2][3].

When the quality functions are all linear, the optimal mix is determined by linear programming methods: the total cost of materials $c = \Sigma q_i v_i$ is to be minimized subject to the linear equation (13) for $x = \bar{x}$ and to (14b), which are linear equations in the v_i for $y_h = \bar{y}_h$ or linear inequalities if upper and lower limits, or both, are specified for the y_h.[4] A model of this kind has a certain resemblance to discontinuous substitution with a finite number of linear activities if (13) is written in the form

$$x = \sum_{k=1}^{m} \lambda_k, \quad v_i = \sum_{k=1}^{m} a_{ik} \lambda_k, \quad \text{where } a_{ik} = \begin{cases} 1 \text{ for } i = k \\ 0 \text{ for } i \neq k, \end{cases}$$

but it is trivial in that the activities are defined by the inputs, the v_i being activity levels, and differs from other linear production models also in that factor substitution is constrained by linear quality requirements.

When some of the quality dimensions which define the quality of the blend depend on its composition but are incapable of objective measurement, such as taste, it is not possible to formulate the corresponding quality requirements in mathematical terms; all we know is that the range of substitution as defined by (13) and by those quality specifications which can be expressed explicitly may not be acceptable in its entirety. However, it is sometimes possible to demarcate the region of acceptable factor combinations by applying some objective criterion which is more or less vaguely correlated with the quality in question (e.g. the contents of certain ingredients as representing taste), or to use a numerical grading scale as an indicator of quality. If these roads are barred, some sort of trial-and-error procedure must be resorted to.

---

[1] This is known in chemical technology as the so-called "Mixture Law" when applicable to physical properties of liquid and solid solutions; cf. PERRY (1941), p. 616.

[2] At least when x and the v_i are measured in units of weight, cf. above.

[3] The ingredients may be thought of as given blends of wholly or partly the same constituents, and the product as a blend of blends whose composition in terms of constituents is a weighted average of the compositions of the respective ingredients. The optimization problem becomes trivial when constituents are themselves used as ingredients (e.g. the manufacturing of vitamin pills from synthetic vitamin preparations rather than from natural foods each containing more than one vitamin).

[4] One of the classic problems of linear programming is the so-called "*diet problem*," where a given selection of foods is "mixed" such as to make a "diet" which satisfies given nutritional requirements at minimum cost, the y_{hi} and y_h being the contents of nutrients (vitamins, calories) per unit of ingredient (food no. i) and per unit of blend (diet) respectively. (Eq. (13) may be disregarded if the total quantity of food per day is not specified.)

Linear blending problems of this kind appear to be of considerable practical importance, especially in the food industry. Some of the first successful practical applications of linear programming were problems of optimal blending with specifications of contents or other qualities which "blend linearly" at least approximately. Examples and references are given in DANØ (1960), Chs. III and V.

(b) In the blending problems above, all other inputs than materials—such as labour and energy required for the mechanical mixing of the ingredients—have been disregarded.

This is permissible when the other inputs do not appear in the quality functions (14) any more than in the material balance (13). For example, they may be uniquely determined by output regardless of product quality—i.e., they are limitational shadow factors which are constant for given x—or they may form a separate group of mutually substitutable inputs which is complementary to the group of substitutable material inputs in (13). As an example of the latter case, let v_1, v_2, and v_3 be materials whose relative proportions determine product quality, y, and let v_4 and v_5 denote two other inputs which can be substituted for each other. Then the model is

$$x = v_1 + v_2 + v_3 = g(v_4, v_5)$$
$$y = y\left(\frac{v_1}{x}, \frac{v_2}{x}, \frac{v_3}{x}\right) \tag{15}$$

and the least-cost factor combination for given x and specified y can be determined for each factor group separately[1]; the optimal blending proportions are the same regardless of the combination of v_4 and v_5 that is applied. (Had v_4 and v_5 been limitational inputs, the equation $x = g(v_4, v_5)$ would have been replaced by $v_4 = a_4 x$, $v_5 = a_5 x$ but the conclusion would be the same.)

However, the assumption that product quality depends on the blending proportions only is not always justified. There are cases where the effect of a change in the blending proportions on product quality can be offset by appropriate adjustments of labour and other factors. Instead of (15), for example, the model may be of the form

$$x = v_1 + v_2 + v_3$$
$$y = y\left(\frac{v_1}{x}, \frac{v_2}{x}, \frac{v_3}{x}, x, v_4, v_5\right) \tag{16}$$

where v_4 and v_5 no longer form a separate factor group but come into the model through the quality function so that we have a model of the type (11)–(12). The optimal combination of materials for given level of output and specified product quality no longer depends on the prices of materials only; the costs of v_4 and v_5 will have to be considered in the cost function which is to be minimized[2].

---

[1] Cf. Ch. V, C.

[2] An empirical example of this type is implicit in the production function for chocolate (FRISCH (1935)) which is cited in Ch. V, A above. The product is made by heating, moulding, and cooling a mixture of two ingredients,

(i) $$x = v_1 + v_2$$

where the amount of processing work, v_3—measured in terms of the quantity of chocolate mix to be processed—is a shadow factor complex depending on the percentage of defective castings in a single moulding operation, h, which in turn depends on the blending proportion v_2/v_1. Let y denote the percentage of perfect castings required in the finished product. Then the second equation of the model is

(ii) $$v_3 \cdot (1-h) = x \cdot y \qquad \text{where } h = h(v_2/v_1)$$

which states that the non-defective units produced per period are $(1-h)$ times the quantity of mix that is processed per period and y times total output per period. The quality parameter y may be specified at any value within the interval $1-h \leq y \leq 1$. When, as in FRISCH's example, no defective units are tolerated—i.e., $y=1$—the defective castings resulting from each operation are reprocessed in subsequent operations so that we have

$$v_3 \cdot (1-h) = x.$$

On the other hand, if all defective units are accepted such as they are (i.e., neither scrapped nor reprocessed) so that $y = 1-h$, a single operation will do so that we have

$$v_3 = x.$$

For y specified at some value between $1-h$ and 1, some but not all of the defective units made in the first moulding operation will have to be reprocessed. For example, let h be $= 0.20$ and $y = 0.95$; then for 100 kg. of paste processed there will be 20 kg. of defective castings, of which 18.75 kg. are reprocessed, leaving only $(20 - 18.75) + 0.20 \times 18.75 = 5$ kg. of defective units.

Eq. (ii) may be written as a quality function

$$y = \frac{[1 - h(v_2/v_1)] \cdot v_3}{x},$$

cf. (16), or as a shadow function with a quality parameter,

$$v_3 = \frac{x \cdot y}{1 - h(v_2/v_1)}.$$

For given output and given blending ratio, v_3 increases with product quality y.

Chapter IX

Plant and Process Production Models

A. Process Interdependencies and Optimization

As stated above[1], the total productive operations that take place within a plant can be broken down into a number of processes each of which contributes to the transformation of raw materials into the final product or products.

For purposes of production analysis it is usually convenient to decompose the problem in this manner—that is, to consider the relationships between the inputs and outputs of individual processes—because the process, being a more fundamental technological unit, is presumably characterized by analytically simpler production relations than is the more complex unit of the plant. The decomposition of plant operations is by no means unambiguous; in loose usage a process may be taken to mean anything from the operation associated with a single unit of fixed equipment (e.g. a machine) to the aggregate of plant operations, and a process may be further decomposed into subprocesses. The process production models developed in the preceding chapters are to be interpreted in this light. In concrete cases, however, some guidance may be found in engineering criteria; the basic technological data which are to be analysed and described in terms of a production model will frequently refer to well-defined individual processes[2].

On the other hand, the problem of optimum operation with which the firm is confronted refers to the plant as a whole. While certain "local" problems of optimum allocation (cost minimization) in a particular process may in some cases be solved independently without reference to other operations within the plant (*"suboptimization"*), it is generally impossible to decompose the problem of maximizing plant profit into a number of separate problems for the individual processes; there is no guarantee that the separate optima are consistent with the overall optimum solution[3], or even that they are consistent with one another. This is because the processes are interdependent in the technical sense that the process models have

---

[1] Ch. II.

[2] For example, most of the physical transformations that take place in the chemical industries are classified in more or less standardized processes known as "unit operations" (evaporation, filtration, extraction, crystallization, etc.), each associated with a particular kind of fixed equipment.

[3] Cf. CHARNES and COOPER (1961), Vol. I, p. 370.

variables in common: the output of one process may appear as an input of another or the processes may have to share the services of some fixed factor.

Clearly this calls for integration of the various process models into a comprehensive model of the operations that take place within the plant[1]. The resulting *plant production model* provides side conditions for the maximization of total profit, and the overall optimum solution together with the process production functions determines the optimum allocation in each of the component processes. It is the purpose of the present chapter to give a brief outline of a theory of plant production functions, their derivation from interdependent process functions, and the relationship of suboptimization to overall profit maximization[2]. Since the number of ways in which processes of various kinds may be combined within a plant is virtually unlimited, no attempt will be made to formulate a general plant production model; only a few—presumably typical—examples will be dealt with.[3]

B. Vertical Integration of Processes

1. A Sequence of Processes with Continuous Substitution

(a) Let us first consider a plant in which the raw material undergoes a series of successive treatments. This means that the plant is organized as a sequence of processes (stages) which are interdependent in that the first process produces an intermediate product which is used as an input in the next process, and so forth until the final product emerges. A simple two-stage example of this kind is illustrated in *Fig. 41*, where v_1 represents the input of raw material in the first process (P_1) and s is the intermediate product; v_2 and v_4 are other variable factors (labour, etc.) whereas v_3 and v_5 represent the services of fixed factors, assuming that each stage or process is characterized by a particular kind of fixed equipment so that there is one capacity limitation to each process, $v_3 \leq \bar{v}_3$ and $v_5 \leq \bar{v}_5$.

An alternative interpretation of Fig. 41 is that s represents some processing factor[4] (e.g. power, steam, etc.) produced in process P_1 whereas raw material is represented by v_4. In this case the material is subjected to one

---

[1] Plant models may be further integrated into production models for multi-plant firms. However, plants which have nothing in common but management (and possibly buildings) are not interdependent in the above sense so that their respective optimization problems may be solved separately.

[2] The relationship between plant and process models has been much neglected in the literature; it is usually disposed of by the tacit assumption that plant and process coincide. CARLSON (1939), p. 11, mentions the problem but considers it to be solved by the use of internal accounting prices, thus begging the question of suboptimization *vs.* overall optimization. CHENERY (1953) has outlined a theory of plant production functions from the point of view of "engineering" models; a more comprehensive treatment is given in CHENERY (1949), pp. 95—100, to which some important points in the following are owed.

[3] Integration of simultaneous, identical processes (where the plant is divisible in space, being composed of a number of identical units of fixed equipment) has been treated in Ch. VII, B above.

[4] Cf. CHENERY (1949), p. 95.

treatment only (P_2), but the problem of integration is formally similar to the case where P_1 represents an initial stage of processing.

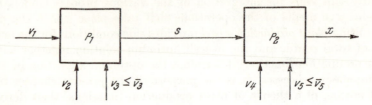

Fig. 41

(b) Assume first that the inputs of each process are continuously substitutable. Then, ignoring for the moment the capacity factors (assuming that they are always fully utilized), the production functions for the two processes have the form

$$s = f(v_1, v_2) \tag{1}$$

$$x = g(s, v_4). \tag{2}$$

The plant production function expressing x in terms of v_1, v_2, and v_4 is readily obtained by eliminating s : [1]

$$x = g[f(v_1, v_2), v_4] = F(v_1, v_2, v_4) \tag{3}$$

so that the integration results in a production function of the same general type as the process functions. The three inputs are continuously substitutable for one another; in the region where the marginal productivities in (1)–(2) are all positive we have

$$F_1' = g_s' \cdot f_1' > 0, \quad F_2' = g_s' \cdot f_2' > 0, \quad F_4' = g_4' > 0$$

so that, for x constant, $\delta v_i / \delta v_j = -F_j'/F_i' < 0$ for any pair of inputs $(i, j = 1, 2, 4)$.

The point of maximum plant profit is determined by maximizing $z = px - q_1 v_1 - q_2 v_2 - q_4 v_4$ subject to the plant production function (3); the first-order maximum conditions

$$p \cdot F_i' - q_i = 0 \qquad (i = 1, 2, 4) \tag{4}$$

together with (3) determine v_1, v_2, v_4, and x.

(4) can be written in the form

$$p = c'(x) = \frac{q_i}{F_i'} \qquad (i = 1, 2, 4)$$

where $c = c(x)$ is the plant cost function, the expansion path along which it is defined being determined by the production function (3) and

$$\frac{q_1}{F_1'} = \frac{q_2}{F_2'} = \frac{q_4}{F_4'},$$

the conditions for minimum cost for parametric x. Because of the constant inputs $\bar{v}_3$ and $\bar{v}_5$ the marginal cost curve may be excepted to be U-shaped.

---

[1] Cf. CHENERY (1949), p. 97, or (1953), p. 311.

2. Problems of Suboptimization

(a) In solving the plant optimization problem we have also determined the optimum allocation in the component processes: for the values of v_1 and v_2 given by (3)–(4), s is determined by the process function (1), or by (2) for the optimal values of x and v_4.

This immediately raises the question whether the two process optima could have been found separately by independent suboptimization of each process.

By the first two equations of (4) we have

$$(p \cdot g_s' =) \frac{q_1}{f_1'} = \frac{q_2}{f_2'}$$

so that the overall optimum satisfies the familiar condition for minimum cost in the first process. In other words, the optimum factor combination in P_1 is a point on the expansion path corresponding to this process, a locus which can be determined without regard to the second process since it depends only on q_1, q_2, and the shape of (1). However, complete suboptimization is not possible without additional information; there is nothing to indicate what particular point on the expansion path represents the optimum level of output in P_1, nor can we determine even the locus of least-cost points belonging to P_2. In order to get any further some sort of price must be attached to the intermediate product s. This is precisely what is done in practical cost accounting[1].

Let the intermediate product be "sold" by the first process to the second at some arbitrarily fixed accounting price, π. Then, maximizing profit in P_1

$$z_1 = \pi s - q_1 v_1 - q_2 v_2$$

subject to (1), we get the necessary conditions

$$\left. \begin{array}{l} \pi \cdot f_1' - q_1 = 0 \\ \pi \cdot f_2' - q_2 = 0 \end{array} \right\} \quad \text{or} \quad \pi = \frac{q_1}{f_1'} = \frac{q_2}{f_2'} \tag{5}$$

which, with (1), determine v_1, v_2, and s. Similarly, the conditions for maximum profit in P_2 as determined by maximizing

$$z_2 = p x - \pi s - q_4 v_4$$

subject to (2) become

$$\left. \begin{array}{l} p \cdot g_s' - \pi = 0 \\ p \cdot g_4' - q_4 = 0 \end{array} \right\} \quad \text{or} \quad p = \frac{\pi}{g_s'} = \frac{q_4}{g_4'} \tag{6}$$

and (2)–(6) determine s, v_4, and x[2].

However, there is nothing to guarantee that the two suboptima will be consistent; indeed, the respective solutions for s will generally differ. The higher the accounting price π is, the more of s will be produced in P_1—cf.

---

[1] Cf. CARLSON (1939), p. 11 n.

[2] (5) and (6) are maximum conditions only if z_1 and z_2 are ≥ 0 at the respective points. If an overall solution exists at which $z > 0$, i.e., $p x - q_4 v_4 > q_1 v_1 + q_2 v_2$, there will exist an interval of π, defined by $p x - q_4 v_4 > \pi s > q_1 v_1 + q_2 v_2$, within which z_1 and z_2 are both positive.

(1)–(5)—and the less of it will be required in P_2 where it is a cost factor, cf. (2)–(6). This result suggests that there is one and only one value of π which will give a consistent solution. Application of any other value of π will lead to suboptimal solutions which are inconsistent with each other.

On the other hand, if consistency is ensured by suboptimizing one process for the value of s determined by suboptimization of the other, the suboptima determined in this way may not satisfy the conditions for overall optimum, (4). For example, let s be determined by (2)–(6), the conditions for maximum profit in P_2, some arbitrary value having been assigned to π. Then the subsequent profit maximization in P_1 reduces to the problem of finding the least-cost combination which will produce the given s so that conditions (5) are replaced by

$$(u =) \frac{q_1}{f_1'} = \frac{q_2}{f_2'} \tag{7}$$

where u is a Lagrange multiplier to be interpreted as marginal cost in P_1. The combined solution (1)–(2), (6)–(7), though consistent, depends on the particular value assigned to π and satisfies the overall optimum conditions (4) only for $\pi = u$;[1] and u in turn generally depends on s, which is one of the unknowns of the problem.

Clearly the dilemma can be resolved by treating π as an unknown in the combined suboptimization problem so that the system (5)–(6) with (1)–(2) is no longer overdetermined. Elimination of π in (5)–(6) is seen to lead to conditions (4), and (1)–(2) combine to give (3). (3)–(4) determine x, s, and the inputs, and π is then determined by either of the equations

$$\pi = p \cdot g_s' = \frac{q_1}{f_1'} = \frac{q_2}{f_2'},$$

that is, the intermediate product is to be priced according to the common value of marginal cost (with respect to s) in P_1 and marginal productivity value of s in P_2. Using this accounting price, a solution is obtained which is consistent and satisfies the overall conditions for maximum plant profit as well as the conditions for suboptimum in each process; no other price will satisfy all these requirements.

However, this means that the overall optimization problem will generally have to be solved before the "correct" accounting price can be fixed, in which case there is little to be gained from suboptimization as a procedure for optimum allocation[2]; the price associated with the intermediate product is useful mainly for pure accounting purposes, the relative distribution of total (gross) profit z between z_1 and z_2 being dependent on the internal price at which the intermediate product is "sold" by one accounting unit

---

[1] This is another illustration of the fact that Lagrange multipliers are often directly interpretable in economic terms.

[2] In practice, however, the accounting price and the optimum allocation can be determined by trial-and-error suboptimization without explicitly solving the analytical problem of maximizing overall profit. Having fixed π at a tentative level—based on an estimate of marginal cost in the first process—and determined the suboptima, it can be examined whether $z_1 + z_2$ will be greater if a higher or lower value of π is applied.

(process) to another. The application of π as an instrument of optimization is helpful only in special cases where the correct accounting price can be determined without first having to solve the overall problem.

This will in fact be possible if one of the process functions is homogeneous of degree one. Suppose that f is a homogeneous function of v_1 and v_2; then marginal cost in P_1 is constant and π can be determined from (5) regardless of (6):

$$\frac{q_1}{f_1'} = \frac{q_2}{f_2'} = \pi$$

where the first equation uniquely determines v_2/v_1 and thus also marginal cost, u, which is independent of s. The constant value of π given by the second equation can then be applied to suboptimization in P_2 and the resulting solution—cf. (6), (2), and (1)—will represent plant optimum.

Conversely, if g is a homogeneous function of s and v_4, the relative factor proportion s/v_4 is constant and independent of x, determined by the second equation of (6); the value of π given by the first equation of (6) is the accounting price to be applied to (5).

Neither of these cases is likely to occur in the present model because there are additional inputs, v_3 and v_5, which are understood to be fixed. However, if these capacity restrictions allow for less than full utilization of capacity, the process production functions may turn out to be homogeneous when the services of the fixed factors are included in the list of variable inputs. We shall now consider the optimization of plant operations under these conditions. In doing so we shall be able to throw some light on optimization subject to several capacity restrictions in inequality form, a kind of problem which has so far been largely evaded by assuming that the process and the production function refer to a single stage of operations characterized by a single capacity factor.

(b) When we drop the assumption that the fixed factors are indivisible, $v_3 = \bar{v}_3$ and $v_5 = \bar{v}_5$, the production functions (1)–(3) will have to be written

$$s = f(v_1, v_2, v_3) \quad (v_3 \leq \bar{v}_3) \tag{1a}$$

$$x = g(s, v_4, v_5) \quad (v_5 \leq \bar{v}_5) \tag{2a}$$

$$x = g[f(v_1, v_2, v_3), v_4, v_5] = F(v_1, v_2, \ldots, v_5) \tag{3a}$$

and the overall optimum allocation is determined by maximizing

$$z = px - q_1 v_1 - q_2 v_2 - q_4 v_4$$

subject to (3a) and to the capacity restrictions $v_3 < \bar{v}_3$, $v_5 \leq \bar{v}_5$.

Side conditions in the form of inequalities always make the optimization problem more difficult to handle analytically[1], the more so when several processes are involved so that there is more than one capacity constraint. All that can be said in general is that the optimum solution is one of the solutions determined by maximizing z (i) subject to (3a) only, (ii)–(iii) sub-

---

[1] The Kuhn-Tucker conditions (cf. Appendix 1, C) give a criterion for maximum under inequalities but do not immediately provide a general method for analytical and numerical solution of problems of this kind.

ject also to either $v_3 = \bar{v}_3$ or $v_5 = \bar{v}_5$, and (iv) subject to both constraints in equality form. If solution (i) exists and is feasible (i.e., turns out to respect both inequalities), it represents the overall optimum; if not, the other three— of which (iv) has been demonstrated above—must be determined and the one which represents the greatest value of z is the best, provided that it is feasible. However, this comparison is possible only when the prices are known and the analytical shapes of the production functions are fully specified.

When the functions f and g, and hence also the plant function F, are homogeneous of degree one—in which case no finite solution to problem (i) exists—a slight short cut may be resorted to. Up to a point the plant expansion path will be linear, the relative factor proportions being uniquely determined by cost minimization for parametric x. Such a path $(0A)$ is shown in a (v_3, v_5) diagram in *Fig. 42*.

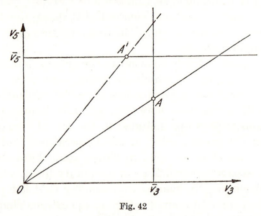

Fig. 42

Expansion along this path will have to stop at point A where v_3 becomes $= \bar{v}_3$.[1] Since the point of maximum profit will have to be somewhere on the rising branch of the marginal cost curve, output will be expanded further and we have to maximize z subject to (3a) and to $v_3 = \bar{v}_3$. The necessary maximum conditions are

$$p \cdot F_i' - q_i = 0 \qquad (i = 1, 2, 4)$$
$$p \cdot F_5' \quad = 0 \tag{4a}$$

which with (3a) for $v_3 = \bar{v}_3$ determine v_1, v_2, v_4, v_5, and x. Provided that it turns out to satisfy $v_5 \leq \bar{v}_5$, the solution represents the overall optimum; if it does not, (3)–(4)—where both capacities are fully utilized—is the optimum solution.

---

[1] The method of finding the bottleneck factor is seen to be similar to that used in the case of limitational inputs with fixed coefficients, only in the present case the position of the linear expansion path depends on the factor prices; for some other set of prices the path would have been $0A'$, in which case $\bar{v}_5$ would have been the effective limit to proportional variation of all inputs. Moreover, in contrast to the limitational model, the capacity of the first process is not a bottleneck in an absolute sense: when the relative factor proportions are capable of variation, further expansion of output is possible through reallocation of the variable inputs after point A has been reached.

The plant cost function is illustrated by the marginal cost curve of *Fig. 43*. The initial phase of constant $c'(x)$—cf. the linear expansion path OA in Fig. 42, as given by

$$(c'=)\frac{q_1}{F_1'}=\frac{q_2}{F_2'}=\frac{q_4}{F_4'}, \quad F_3'=F_5'=0$$

which determine the four relative factor proportions—is followed by a phase of increasing marginal cost (AB) where v_3 is fully utilized, the expansion path being determined by (3a) and the conditions

$$(c'=)\frac{q_1}{F_1'}=\frac{q_2}{F_2'}=\frac{q_4}{F_4'}, \quad v_3=\bar{v}_3, \quad F_5'=0.$$

Solution (3a)–(4a) represents an optimum on this branch of the marginal cost curve: for $p=\bar{\bar{p}}$ (point P in Fig. 43), the optimum condition $p=c'$ is seen to be equivalent to (4a).

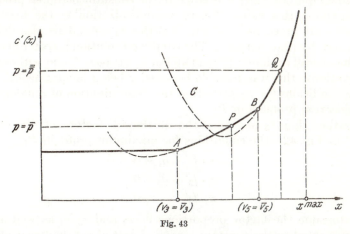

Fig. 43

For increasing values of the parameter p, the higher level of output determined by $p=c'(x)$ will require more of each input, including v_5, up to the point (B) where v_5 becomes equal to $\bar{v}_5$. Further expansion will have to follow the path given by (3a) and

$$(c'=)\frac{q_1}{F_1'}=\frac{q_2}{F_2'}=\frac{q_4}{F_4'}, \quad v_3=\bar{v}_3, \quad v_5=\bar{v}_5,$$

where $c' \to \infty$ as output approaches the maximum capacity of the plant, $x=x^{\max}$ (as determined by $F_i'=0$ $(i=1,2,4)$, $v_3=\bar{v}_3$, $v_5=\bar{v}_5$). Optimum solutions in this third phase of the cost function—cf. point Q where $\bar{\bar{p}}=c'$— are seen to be represented by (3)–(4).

In other words, the marginal cost function is pieced together from three different sections. The composite function is continuous (but not differentiable) at A and B, c' being $=q_1/F_1'$ everywhere. (Had $\bar{v}_3$ and $\bar{v}_5$ been indivisible, as assumed in (1)–(3), the first two phases would have been replaced by the dotted curve CB, the marginal cost curve CBQ being U-shaped; cf. above.)

Optimum solutions on the curve section AB, where $v_3 = \bar{v}_3$ but $v_5 < \bar{v}_5$, can be determined by suboptimization because the process function g is homogeneous in the variable inputs s, v_4 and v_5. The conditions for maximum profit in the respective processes are

$$\pi \cdot f_1' - q_1 = 0$$
$$\pi \cdot f_2' - q_2 = 0 \qquad\qquad (5\,\mathrm{a})$$

(where $v_3 = \bar{v}_3$) and

$$p \cdot g_s' - \pi = 0$$
$$p \cdot g_4' - q_4 = 0 \qquad\qquad (6\,\mathrm{a})$$
$$p \cdot g_5' = 0$$

which are consistent if π is treated as an unknown. Since the partial derivatives of g are functions of the relative factor proportions s/v_4 and s/v_5, the latter are determined uniquely by the second and third equation of (6a) and the first equation gives π as equal to the constant marginal productivity value. Applying this value of π to suboptimization in the first process, v_1, v_2, and s are given by (5a) and (1a) (for $v_3 = \bar{v}_3$). This also determines v_4 and v_5 and, by (2a), output x. The combined solution is seen to be equivalent to the overall optimum (3a)–(4a), but it has been found without solving these equations; this was possible because, g being homogeneous, the suboptimum conditions for P_2 permitted the determination of π independently of the allocation problem in P_1.

If $\bar{v}_5$ rather than $\bar{v}_3$ had been the bottleneck factor—cf. the expansion path OA' in Fig. 42—the conditions for maximum profit in P_1,

$$\pi \cdot f_1' - q_1 = 0$$
$$\pi \cdot f_2' - q_2 = 0 \qquad\qquad (5\,\mathrm{b})$$
$$\pi \cdot f_3' = 0$$

would determine the factor proportions v_1/v_3 and v_2/v_3 as well as π, the latter being equal to the constant marginal cost along the linear expansion path. Using this accounting price in the conditions for suboptimum in P_2,

$$p \cdot g_s' - \pi = 0$$
$$p \cdot g_4' - q_4 = 0 \qquad\qquad (6\,\mathrm{b})$$

(where $v_5 = \bar{v}_5$), we get a solution (1a)–(2a), (5b)–(6b), which coincides with the overall optimum solution determined by (3a) and the conditions for maximum plant profit,

$$p \cdot F_i' - q_i = 0 \qquad (i = 1, 2, 4)$$
$$p \cdot F_3' = 0. \qquad\qquad (4\,\mathrm{b})$$

3. Other Examples of Vertical Integration

(a) Now let us assume that the intermediate product produced by the first process is a limitational shadow factor in the second, all other inputs being substitutable in the respective processes as before. Then the process models can be written

$$s = f(v_1, v_2, v_3) \qquad (v_3 \leq \bar{v}_3) \tag{8}$$

and

$$x = g(v_4, v_5) \qquad (v_5 \leq \bar{v}_5)$$
$$s = a \cdot x \tag{9}$$

where a is the constant coefficient of production. The assumption that s is proportional to x is probably realistic in many cases where s is a material (produced in P_1) to be further processed in P_2.

In cases of this type elimination of s leads to an integrated plant model consisting of two equations[1],

$$x = \frac{1}{a} \cdot f(v_1, v_2, v_3) = g(v_4, v_5) \qquad (v_3 \leq \bar{v}_3, \ v_5 \leq \bar{v}_5), \tag{10}$$

one for each process[2].

The plant optimum is determined by maximizing total profit subject to the two production functions (10) and to the corresponding capacity constraints. Again, if the functions f and g are assumed to be homogeneous of the first degree, the bottleneck factor can be detected by confronting the linear phase of the expansion path in a (v_3, v_5) diagram with the limitations $v_3 \leq \bar{v}_3$ and $v_5 \leq \bar{v}_5$. Let v_3 be the effective limit; then, maximizing the Lagrangian

$$L = p \cdot \frac{1}{a} \cdot f(v_1, v_2, \bar{v}_3) - q_1 v_1 - q_2 v_2 - q_4 v_4 + u \cdot \left(g(v_4, v_5) - \frac{1}{a} \cdot f(v_1, v_2, \bar{v}_3) \right)$$

and eliminating the multiplier u, we get the conditions

$$\frac{1}{a} \cdot \left(p - \frac{q_4}{g_4'} \right) = \frac{q_1}{f_1'} = \frac{q_2}{f_2'} \tag{11}$$
$$g_5' = 0.$$

The solution determined by (10)–(11) with $v_3 = \bar{v}_3$ represents plant optimum[3]. The first two equations of (11) are to be interpreted as an equality between the marginal cost of producing s in P_1 and the marginal profit derived from processing another unit of s in P_2, $\delta z_2 / \delta s$.

Suboptimization in P_1, i.e., maximization of

$$z_1 = \pi \cdot f(v_1, v_2, \bar{v}_3) - q_1 v_1 - q_2 v_2,$$

leads to the conditions

$$\pi \cdot f_1' - q_1 = 0$$
$$\pi \cdot f_2' - q_2 = 0 ; \tag{12}$$

the conditions for maximum of

$$z_2 = px - \pi s - q_4 v_4 = (p - \pi a) \cdot g(v_4, v_5) - q_4 v_4$$

---

[1] Cf. CHENERY (1949), p. 97.

[2] This is an example of complementary groups of substitutional inputs or, in FRISCH's terminology, "disconnected factor rings," cf. Ch. V, C.

[3] Unless v_5 turns out to be greater than $\bar{v}_5$, in which case z must be maximized over again for $v_5 = \bar{v}_5$ as well as $v_3 = \bar{v}_3$; the former condition then replaces $g_5' = 0$ in (11).

are

$$(p - \pi a) \cdot g_4' - q_4 = 0$$
$$(p - \pi a) \cdot g_5' \qquad = 0.$$

(13)

Assuming that g is a homogeneous function, (13) determine the relative factor combination v_4/v_5 and π, the latter being equal to

$$\pi = \frac{1}{a} \cdot \left(p - \frac{q_4}{g_4'} \right)$$

which is constant, independent of output. Substitution of this value in (12) is seen to lead to the plant optimum conditions (11). In other words, provided that the function g is homogeneous, the complete problem can be solved by suboptimization, starting with the second process.

Had $\bar{v}_5$ turned out to be the bottleneck capacity, the conditions for maximum profit in the respective processes would have been

$$\pi \cdot f_1' - q_1 = 0$$
$$\pi \cdot f_2' - q_2 = 0$$
$$\pi \cdot f_3' \qquad = 0$$

(12 a)

which for f homogeneous of degree one leads to a uniquely determined accounting price equal to the constant marginal cost (= average variable cost) in P_1, and

$$(p - \pi a) \cdot g_4' - q_4 = 0$$

(13 a)

(where $v_5 = \bar{v}_5$). For the value of π given by (12 a) the combined solution (8)–(9), (12 a)–(13 a) (for $v_5 = \bar{v}_5$) is equivalent to the overall maximum of z as determined by (10) and the conditions

$$\frac{1}{a} \cdot \left(p - \frac{q_4}{g_4'} \right) = \frac{q_1}{f_1'} = \frac{q_2}{f_2'}$$

(11 a)

$$f_3' = 0. \text{ [1]}$$

(b) Suppose instead that the inputs of each process are all limitational,

$$v_i = a_i \cdot s \qquad (i = 1, 2, 3), \quad v_3 \le \bar{v}_3$$

(14)

---

[1] It must be emphasized that suboptimization is impossible when the production functions for the successive processes have other variables than the intermediate product in common.

FRISCH's chocolate production function—cf. Ch. V, A above—is a case in point. The production function may be decomposed into process functions for mixing and moulding-cooling respectively:

$$(P_1) \qquad s = v_1 + v_2$$

and

$$(P_2) \qquad x = v_3 \cdot [1 - h(v_2/v_1)]$$
$$s = x$$

where s is the quantity of chocolate mix to be moulded and cooled. Though in some respects similar to (8)–(9), the model differs in that v_1 and v_2 appear in both process functions: the amount of moulding-cooling work to be done depends on the liquidity of the mix, which in turn depends on the mixing proportion. The least-cost mix, therefore, cannot be determined independently of the optimization problem in the second process and vice versa; a mix consisting only of the cheaper input (v_1) is not consistent with the overall cost minimum.

and

$$s = a_s \cdot x$$
$$v_i = a_i \cdot x \qquad (i = 4, 5), \qquad v_5 \leq \bar{v}_5$$

(15)

where the a_i are constant coefficients of production[1]. (14)–(15) immediately reduce to an integrated plant production model of the same type,

$$v_i = (a_i a_s) \cdot x \qquad (i = 1, 2, 3)$$
$$v_i = a_i \cdot x \qquad (i = 4, 5).$$

(16)

Total variable plant cost is proportional to output, and plant optimum—provided that unit gross profit is ≥ 0—is determined by the bottleneck capacity,

$$x = \text{Min} \left(\frac{\bar{v}_3}{a_3 a_s}, \frac{\bar{v}_5}{a_5} \right). \quad [2]$$

(17)

Linear constant-coefficient models with several capacity limitations are usually to be interpreted in this way, being derived from—or decomposable into—models of successive processes (stages) each of which is associated with a particular fixed factor[3] [4].

---

[1] If s is measured in units of x, a_s will be $= 1$.

[2] The shadow price associated with the effective bottleneck factor—cf. Ch. III, A—expresses the increase in profit made possible by a unit increase in the factor's capacity, i.e., the marginal profitability of investment with a view to harmonizing the capacities of the processes.

[3] FRENCKNER (1957), pp. 81 ff., has given an empirical example of a model of this kind. A machine shop turns out three different products (two kinds of lathes and a mixing machine) using the same fixed factors, and each product is made in a single activity with constant coefficients at the respective stages. There are five successive stages, each defined by an operation performed on a particular kind of fixed equipment—two kinds of lathe operations, milling, drilling, and assembly—and therefore five capacity limits.

The sales potential is virtually unlimited so that the firm can sell as much as it can turn out if any one of the products is produced exclusively; however, since this would result in idle capacity at those stages which are not effective bottlenecks, it is more profitable to make several products jointly.

[4] JANTZEN's "Law of Harmony"—cf. JANTZEN (1924), pp. 39 ff., or (1939), pp. 33 ff.; see also BREMS (1952 a, b)—is based on a production model of this type, only the capacity factors are treated as variable since JANTZEN was concerned with the behaviour of total unit cost when the capacities of the processes (stages) are adjusted along with the expansion of output in such a way as to minimize total cost for any level of output. (JANTZEN's total plant cost per period is the sum of variable (operating) cost, which is proportional to output, and "fixed" (or "equipment") cost per period, the latter being defined for each kind of durable factor as the value of the capital stock divided by its useful life in terms of periods.)

While the services of fixed equipment are perfectly divisible in the time dimension, the units in which these durable factors as such are purchased are large and indivisible so that the capacities of the successive processes—cf. (17)—are usually not geared to each other; in JANTZEN's terminology, the composition of the plant is disharmonious. In the example above, let $\bar{v}_3$ be the effective limit; then output can be expanded further only by installing another unit of the third factor, thus doubling the capacity of the first stage; and so forth. Whenever a new unit of durable equipment is purchased and installed, total plant cost per unit of output jumps in a discontinuous manner; unit cost will be a minimum only for those levels of output which are common multiples of the capacities (in terms of x) of the respective units of the fixed factors. At such a point the plant is harmoniously composed and idle capacity does not occur at any stage.

(c) Finally, let each of the process models be linear and characterized by discontinuous substitution:

$$s = \sum_{j=1}^{M} \mu_j$$

$$v_i = \sum_{j=1}^{M} a_{ij}\mu_j \qquad (i=1, 2, 3), \quad v_3 \leq \bar{v}_3 \tag{18}$$

and

$$x = \sum_{k=1}^{N} \lambda_k$$

$$s = \sum_{k=1}^{N} a_{sk}\lambda_k \tag{19}$$

$$v_i = \sum_{k=1}^{N} a_{ik}\lambda_k \qquad (i=4, 5), \quad v_5 \leq \bar{v}_5.$$

μ_j and λ_k are the activity levels (required to be ≥ 0) of the sub-activities in P_1 and P_2 respectively. Eliminating s, we get a linear plant model where x and some of the v_i are linear functions of the λ_k and the other inputs depend on the μ_j, the two groups of activity levels being connected by the linear equation

$$(s=) \sum_{j=1}^{M}\mu_j = \sum_{k=1}^{N} a_{sk}\lambda_k. \tag{20}$$

The optimum allocation of the plant is determined by maximizing total profit, which can be expressed as a linear function of the $M + N$ activity levels, subject to (20) and to the capacity limitations $v_3 \leq \bar{v}_3$, $v_5 \leq \bar{v}_5$ which are linear inequalities in the μ_j and the λ_k respectively[1]. At most three activity levels will be positive in the optimum solution.

Model (18)–(19) assumes that the intermediate product turned out by P_1 is the same no matter how it is made and can be used as an input for any method of production applied in P_2.[2] Now assume instead that P_1 and P_2 are technically interdependent in the stronger sense that a unit of final output produced by activity no. k in P_2 requires a constant amount a_{sk} of the intermediate product s_k turned out by the k'th activity of the first stage; any other activity in P_1 produces a technically different intermediate product which does not fit in. This means that the second equation of (19) will have to be replaced by

$$s_k = a_{sk}\lambda_k$$

---

[1] Linear programming problems of this type are particularly easy to solve numerically because of the many zeroes in the coefficient matrix.

[2] This assumption is implicit in the first equation of (18) and the second equation of (19): the μ_j as well as the $a_{sk}\lambda_k$ are additive.

(where a_{sk} may often for convenience be set $= 1$) and the model for the first process becomes

$$v_i = \sum_{k=1}^{N} a_{ik} s_k \qquad (i = 1, 2, 3).$$

The integrated plant model becomes

$$x = \sum_{k=1}^{N} \lambda_k$$

$$v_i = \sum_{k=1}^{N} a_{ik} a_{sk} \lambda_k \qquad (i = 1, 2, 3), \qquad v_3 \leq \bar{v}_3 \qquad (21)$$

$$v_i = \sum_{k=1}^{N} a_{ik} \lambda_k \qquad (i = 4, 5), \qquad v_5 \leq \bar{v}_5,$$

which is a linear discontinuous-substitution model with N activities, each of which corresponds to a particular technical quality of the intermediate product. Only two activities (at most) will be used in the optimum allocation[1], corresponding to the number of capacity constraints[2].

---

[1] In solving the plant optimization problem by linear programming, shadow prices of the fixed factors are implicitly determined, cf. Ch. III, B. These imputed prices represent the marginal values to the firm of additional capacity in the respective processes and can therefore be used as a guide to investment decisions with a view to increasing the capacities.

[2] CHARNES, COOPER and FARR (1953) have given an empirical example of a linear plant model which may be interpreted as a kind of intermediate case between (18)–(19) and (21). The plant turns out a multitude of products which are scheduled through the same sequence of mechanical processes, the more important capacity factors of which are screw machines and grinders respectively. Each product can be produced in a number of linear activities.

Consider one such product (no. 5) in isolation. There are four activities, all of which have the same input coefficient for each variable factor so that unit profit is the same in all four activities; they differ only in respect to the input coefficients of the capacity factors. The matrix of these coefficients (machine times per unit) has the form (op. cit., pp. 120 f.):

		λ_1	λ_2	λ_3	λ_4	
P_1	v_{1a}	a_{1a}	a_{1a}			$\leq \bar{v}_{1a}$
	v_{1b}			a_{1b}	a_{1b}	$\leq \bar{v}_{1b}$
P_2	v_{2a}	a_{2a}		a_{2a}		$\leq \bar{v}_{2a}$
	v_{2b}		a_{2b}		a_{2b}	$\leq \bar{v}_{2b}$

The interpretation of the underlying model appears to be as follows:

The first processing stage (P_1) can be performed by either of two types of screw machines, with capacities $\bar{v}_{1a}$ and $\bar{v}_{1b}$ respectively; similarly there are two types of grinders capable of doing each other's jobs. Each of the four machine types defines a sub-activity, characterized by a machine-time coefficient (a_{1a}, etc.). The intermediate products produced by the two sub-activities at the first stage (P_1) are apparently not technically identical—if they were they would be additive as in (18)–(19)—but either one can be used as an input for either sub-activity at the subsequent stage (P_2) so that there will be four plant activities as shown in the table. (In general, there will be $M \cdot N$ plant activities as against $M + N$ in the plant model corresponding to (18)–(19), where each sub-activity in P_2 could be combined with any *combination* of sub-activities in P_1 producing the requisite amount of intermediate product.)

The "horizontal" division of the plant model into parallel activities, as expressed in (21), is more convenient for purposes of optimization than the vertical division into successive stages (processes associated with particular kinds of fixed equipment), not only because a straightforward computational method suited to (21) is available, but also because the close technical interdependency between the stages precludes process suboptimization at any level. Whereas in model (18)–(19) the least-cost combination of the first three inputs—those associated with P_1—could be determined for any given s independently of the λ_k, it is characteristic of model (21) that the optimal choice of activities to be used at the first stage will depend on the prices of the inputs used in the second process because the horizontal division into activities cuts all the way through the plant[1].

C. Integration of Parallel Sequences of Processes

(a) Now assume, as a generalization of the previous models, that the final stage of processing requires two intermediate products, s_1 and s_2, each produced in a separate process[2]. (Alternatively s_1 may be a processing factor.) In this case there will be three processes to be integrated as illustrated by *Fig. 44*. Each process is assumed to require the services of a single capacity factor.

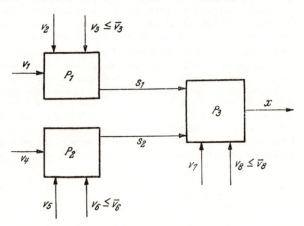

Fig. 44

Any plant whose product is made by assembling parts produced separately within the plant can be illustrated in this way.

(b) Assuming that each process is characterized by continuous factor substitutability, the process production functions have the form

---

[1] Cf. the analogous observation above on Frisch's chocolate production function, where the optimal combination of raw materials in the mixing process could not be determined independently of the optimization at the subsequent stage (moulding-cooling).

[2] In Chenery's terminology, the processes which produce s_1 and s_2 are joint with respect to output; cf. Chenery (1949), p. 96.

$$s_1 = f(v_1, v_2, v_3) \qquad (v_3 \leq \bar{v}_3) \qquad (22)$$

$$s_2 = g(v_4, v_5, v_6) \qquad (v_6 \leq \bar{v}_6) \qquad (23)$$

$$x = h(s_1, s_2, v_7, v_8) \qquad (v_8 \leq \bar{v}_8). \qquad (24)$$

The integrated production function for the plant becomes

$$x = h(f, g, v_7, v_8) = F(v_1, \ldots, v_8) \qquad (v_3 \leq \bar{v}_3, \ v_6 \leq \bar{v}_6, \ v_8 \leq \bar{v}_8) \qquad (25)$$

where all inputs are substitutable.

The plant expansion path, as determined by minimizing total cost for parametric x, and the corresponding cost function will pass through four phases. For sufficiently small x there will be idle capacity in all processes. At a certain level of output one of the three capacity factors becomes fully utilized; from some later point, two capacities will be exhausted; and so on. Each time a capacity limit is reached, the equation $F_i' = 0$ for the factor in question is replaced by $v_i = \bar{v}_i$ in the equilibrium conditions; the marginal cost function will be continuous but kinked at these points[1].

Again, the allocation which leads to maximum plant profit can be determined by suboptimization only in special cases where marginal cost is constant in one of the processes. Assume that the three process production functions and thus also (25) are homogeneous of degree one so that the first phase is characterized by constant marginal cost along a linear expansion path. Let $v_3 \leq \bar{v}_3$ be the effective capacity limit which puts a stop to expansion along this path[2]; then we know that $v_3 = \bar{v}_3$ at the point of overall optimum.

As long as the capacity factor of the second process is less than fully utilized, marginal cost in P_2, dc_2/ds_2, will be constant, the locus of least-cost points being determined by (23) and

$$\left(\frac{dc_2}{ds_2}=\right) \frac{q_4}{g_4'} = \frac{q_5}{g_5'}, \quad g_6' = 0. \qquad (26)$$

Applying the constant marginal cost as the accounting price π_2 of s_2,

$$\pi_2 = \frac{dc_2}{ds_2}, \qquad (27)$$

the conditions for suboptimal allocation in the second process are seen to be satisfied. The conditions for maximum profit in the third process—assuming that there is sufficient capacity—are

$$
\begin{aligned}
p \cdot h_f' - \pi_1 &= 0 \\
p \cdot h_g' - \pi_2 &= 0 \\
p \cdot h_7' - q_7 &= 0 \\
p \cdot h_8' &= 0.
\end{aligned}
\qquad (28)
$$

---

[1] Cf. the three-phase cost function (corresponding to a model with two capacity factors) shown in Fig. 43.

[2] For given prices the bottleneck is determined in a similar way to that applied in Fig. 42, only there are now three capacity restrictions.

For the given value of π_2, the three relative factor proportions in P_3 are determined by the last three equations of (28); if π_1 is set $= p \cdot h_f'$—i.e., if the first intermediate product is priced according to its constant marginal productivity value in P_3—the complete conditions for maximum profit in P_3 are satisfied. Finally, applying this value of π_1 to the conditions for maximum profit in P_1,

$$\pi_1 \cdot f_1' - q_1 = 0$$
$$\pi_1 \cdot f_2' - q_2 = 0 \tag{29}$$

(where $v_3 = \bar{v}_3$), the three suboptima given by (26)–(29) with (22)–(24) are seen to form a consistent solution which is equivalent to the overall optimum solution as determined by (25) and the conditions

$$p \cdot F_i' - q_i = 0 \qquad (i = 1, 2, 4, 5, 7)$$
$$p \cdot F_i' \quad\;\; = 0 \qquad (i = 6, 8) \tag{30}$$

for $v_3 = \bar{v}_3$. If it is feasible—i.e., if it respects the capacity limitations for P_2 and P_3—the solution represents the overall optimum; if it is not, a new solution will have to be found, setting $v_6 = \bar{v}_6$ or $v_8 = \bar{v}_8$ from the outset.

(c) When s_1 and s_2 are interpreted as parts to be assembled in the third process, it is perhaps more realistic to assume that they are limitational inputs in P_3.[1] Then (24) has to be replaced by

$$x = h(v_7, v_8)$$
$$s_1 = a_1 \cdot x \tag{24a}$$
$$s_2 = a_2 \cdot x$$

and (25) becomes

$$x = \frac{1}{a_1} \cdot f(v_1, v_2, v_3) = \frac{1}{a_2} \cdot g(v_4, v_5, v_6) = h(v_7, v_8) \tag{25a}$$

which is characterized by complementary groups of substitutional inputs, each group corresponding to a process[2].

Still assuming that the functions are homogeneous of the first degree and that v_3 is the bottleneck factor, maximize the Lagrangian

$$L = p \cdot h - \sum_i q_i v_i + u_1 \cdot (f - a_1 \cdot h) + u_2 \cdot (g - a_2 \cdot h)$$

for $v_3 = \bar{v}_3$. This leads to the conditions

---

[1] FRISCH (1953), p. 6, exemplifies this by a cutlery works: one blade and one handle are required to produce a knife.

[2] Cf. Ch. V, C above. FRISCH (1953), pp. 6 f., derives the plant production function for a general case of this type in a slightly different way, treating the limitational factors s_1, s_2, ... as *independent* variables so that the plant function becomes

$$x = \text{Min}(s_j / a_j)$$

(cf. Ch. III, A, model (1c)). This of course makes no difference to the optimum solution.

$$u_1 = \frac{q_1}{f_1'} = \frac{q_2}{f_2'}, \quad v_3 = \bar{v}_3$$

$$u_2 = \frac{q_4}{g_4'} = \frac{q_5}{g_5'}, \quad g_6' = 0 \tag{30a}$$

$$p = u_1 a_1 + u_2 a_2 + \frac{q_7}{h_7'}, \quad h_8' = 0.$$

The solution determined by (30a) and (25a) represents the plant optimum provided that it respects the other two capacity constraints. Suboptimization clearly leads to the same result for $\pi_1 = u_1$ and $\pi_2 = u_2$, where u_1 and u_2 can be interpreted as marginal cost in P_1 and P_2; u_2 is a constant determined independently by the second group of equations (i. e., the conditions for maximum profit in P_2) and $\pi_1 = u_1$ then follows from the last equations, h_7' and h_8' being functions of v_7/v_8.

(d) The models above are examples of "parallel" processes which are interdependent through the joint application of their products at a subsequent stage of processing. Another kind of interdependency results from the joint use of some fixed factor such as a machine or a processing factor subject to a capacity limitation[1]. The fact that labour and other current inputs are used in both processes does not in itself establish any relationship between them[2]—the amounts of labour required by the respective processes can be varied independently—but if the processes have to share the services of the plant's fixed equipment the processes cannot be considered separately.

This kind of interdependency is particularly important in multi-product plants where the several products are processed by wholly or in part the same machines, a case which will be dealt with in the following chapter.

---

[1] In CHENERY's terminology, the processes are joint with respect to inputs, cf. CHENERY (1949), p. 96.

[2] CHENERY, *op. cit.*, pp. 97 f. The joint use of buildings can generally be disregarded since the available floor space does not represent an effective bottleneck in the short run, cf. *op. cit.*, p. 99.

Chapter X

Multi-Product Models

A. Alternative Processes and Joint Production

We shall now proceed to the more general case of multiple production, where two or more products are manufactured by the same plant. Different grades or qualities of the "same" product will be regarded as so many distinct products.

In view of the fact that multiple production appears to be the rule rather than the exception, it may well be said that single-output models have been given rather more than due attention in the literature. To some extent, perhaps, this is to be explained by the more complex nature of input-output relationships in multi-product cases. However, some justification of the predominance of single-output models in the theory of production may be sought in the fact that it is often possible to decompose a multi-product model into separate models for the respective products, not only when the products are manufactured in separate plants under common management—in cases of this type the models for each product have no explicit variables in common—but also when they are made in distinct processes which have only the joint use of the plant's fixed facilities in common[1]. In the latter case it is still possible to formulate a separate technological production model for each output, only the models are economically interrelated and must be integrated for purposes of plant optimization[2] because the processes—i.e., the products—become *alternative* as soon as a common capacity factor is fully utilized.

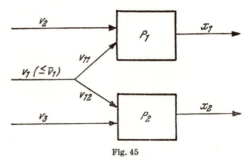

Fig. 45

[1] FRISCH (1956), p. 7.
[2] Cf. Ch. IX above.

A simple case of this type is illustrated in *Fig. 45*, where the processes P_1 and P_2 share the limited capacity of a fixed factor's services, $v_1 \leq \bar{v}_1$; for example, the same machine is used in manufacturing the two products. The amounts of the respective inputs that go into each process can be identified (e.g. v_{11} hours of machine service are used in producing x_1 units of the first output and v_{12} units for x_2); the technical relationships between v_{11}, v_2, and x_1 represent the production model for P_1, and similarly for P_2. The two process models are interdependent in that $v_1 = v_{11} + v_{12}$ cannot exceed the capacity limit, $\bar{v}_1$.

However, in a multitude of cases such a simple decomposition is not possible, because the processes are technically interdependent also in other respects (having other variables than plant services in common)[1] or because the products are made *jointly* in the same process which cannot be technically divided into subprocesses for the respective products; as an extreme case, the products are manufactured in technically fixed proportions. In neither case is it possible to allocate all inputs between the outputs, and the production model will have to refer to a non-decomposable process representing truly joint production. This is illustrated in *Fig. 46* below.

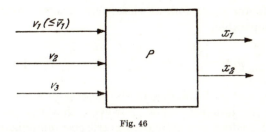

Fig. 46

It is possible, of course, to treat the case of Fig. 45 as a single process producing two outputs with three inputs (v_1, v_2, v_3); however, since it is analytically convenient to decompose the model whenever possible, such cases will be treated as *alternative processes* and only the non-decomposable models will be referred to as cases of truly *joint production* (*multi-product processes*)[2].

In the following we shall examine a number of presumably typical multi-product models with special reference to the possibilities of product and factor substitution.

---

[1] Cf. the analogous observation in Ch. IX, B above.

[2] This fundamental distinction will usually but not always coincide with the criterion whether or not it is possible (though not necessarily economical) to produce each output without making any of the others. Limiting cases are conceivable where solutions on the boundary of the range of product substitution are possible even though the joint process cannot be decomposed.

A production process which is technically a case of joint production may from the point of view of economic analysis represent single production, namely when only one of the goods produced can be sold at a positive price. Under different market conditions the case may come to represent multiple production in an economic sense.

B. Alternative Processes

1. Process Additivity

Let n different products be produced simultaneously in a given plant but in separate processes[1] $P_1, P_2, \ldots, P_n$. Process no. j turns out x_j units per period of the j'th product, using v_{ij} units of the i'th input ($i = 1, 2, \ldots, m$). Assume further that the processes are interdependent only in the sense that they share the services of the plant's fixed equipment[2]. Each process has a production function of its own, relating the v_{ij} to x_j, and the processes are additive factor for factor,

$$v_i = \sum_{j=1}^{n} v_{ij} \tag{1}$$

where v_i is the total amount of input no. i consumed per period; the only link between the P_j is a joint capacity restriction

$$v_s = \sum_{j=1}^{n} v_{sj} \leq \bar{v}_s \tag{2}$$

for each fixed factor v_s. It follows that total (plant) profit and total cost are additively separable,

$$z = \sum_{j=1}^{n} p_j x_j - \sum_{i=1}^{m} q_i v_i = \sum_{j=1}^{n} \left(p_j x_j - \sum_{i=1}^{m} q_i v_{ij} \right) = \sum_{j=1}^{n} z_j \tag{3}$$

for any set of x_j and v_{ij} belonging to the process production functions and satisfying (2)[3]. (It does *not* follow, however, that the *plant cost function* $c = c(x_1, \ldots, x_n)$, as determined by parametric cost minimization, can always be written $c = c_1(x_1) + \ldots + c_n(x_n)$.)

Fig. 45 above illustrates a simple example of this type where $v_1 \leq \bar{v}_1$ is the joint capacity restriction; the variable inputs v_2 and v_3 may be qualitatively different inputs, each used in one process only, or they may represent the amounts used of the same variable input (for example, labour or a homogeneous complex of labour and raw material)[4].

2. Linear Models

(a) Let us first assume that each process is characterized by *constant coefficients of production* (limitational inputs),

---

[1] Or sequences of sub-processes representing successive stages of manufacturing the product in question (cf. Ch. IX, B above).

[2] Cf. CHENERY (1949), p. 96, and Ch. IX, C above.

[3] We are here concerned with variable costs only. The allocation of fixed cost elements, which remains arbitrary, is of no consequence in short-run optimization models.

[4] In the latter case v_2 and v_3 might be replaced by the symbols v_{21} and v_{22}. However, it makes no difference whether or not they represent the "same" factor since there is no restriction on the total quantity of an input which can be purchased in unlimited amounts on the market.

$$v_{ij} = a_{ij} \cdot x_j \qquad (i = 1, 2, \ldots, m; \; j = 1, 2, \ldots, n) \qquad (4)$$

so that the multi-product model for the plant has the form

$$v_i = \sum_{j=1}^{n} a_{ij} x_j \qquad (i = 1, 2, \ldots, m) \qquad (5)$$

where the x_j can be varied within the limits set by the capacity constraints $v_s \leq \bar{v}_s$ and the non-negativity requirements $x_j \geq 0$.[1] Maximizing some x_j subject to these restrictions we get the maximum capacity of the plant in this particular direction for given values of the other $n - 1$ x_j; this defines the *capacity curve* (more generally, the capacity surface) of the plant as the boundary of the region of feasible output combinations in product space.

In the two-product case with one capacity factor ($v_1 \leq \bar{v}_1$) the capacity curve obviously is the straight line

$$(v_1 =) \; a_{11} x_1 + a_{12} x_2 = \bar{v}_1,$$

cf. AB in *Fig. 47* where $0AB$ is the region of feasible combinations satisfying $v_1 \leq \bar{v}_1$, $x_1 \geq 0$, $x_2 \geq 0$. On the boundary the two processes are alternative in the sense that one output can be increased only at the expense of the other, $dx_2 / dx_1 = -a_{11}/a_{12} < 0$. The optimum allocation, as determined by maximizing profit—which is a linear function of x_1 and x_2—subject to the linear restriction

$$(v_1 =) \; a_{11} x_1 + a_{12} x_2 \leq \bar{v}_1$$

for $x_1, x_2 \geq 0$, is obviously represented by either A or B (or, as a special case, any point of AB) so that it pays the firm to specialize in one of the products to the exclusion of the other; there is no advantage in producing both outputs simultaneously.

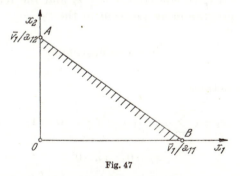

Fig. 47

In the case of two or more capacity restrictions the capacity curve will generally be kinked. Each fixed factor defines a capacity line $v_s = \bar{v}_s$ and thus an upper bound to x_2 for given $x_1 = \bar{x}_1$,

[1] Empirical examples of this type of multi-product model are given by FRENCKNER (1957), pp. 81 ff. (cf. Ch. III, A and IX, B), and by BARGONI, GIARDINA and RICOSSA (1954); the capacity factors in the latter example (cf. Ch. III, A above) are chemical raw materials produced by two supplementary plants with limited capacities.

$$x_2 \leq \frac{\bar{v}_s - a_{s1}\,\bar{x}_1}{a_{s2}},$$

and the lowest upper bound for the given $\bar{x}_1$ represents the maximum of x_2; at the points where the capacity lines intersect in the positive quadrant one capacity factor replaces another as the effective bottleneck. In *Fig. 48*, $0CPB$ is the feasible region where all three capacity restrictions are satisfied; $v_2 \leq \bar{v}_2$ represents the effective limit to x_2 for values of $\bar{x}_1$ up to the critical value $\bar{x}_1^P$, after which $\bar{v}_1$ becomes the bottleneck. The third capacity limit, $v_3 \leq \bar{v}_3$, is automatically satisfied by all points in $0CPB$ and never becomes effective. Hence the broken line CPB is the capacity curve, along which dx_2/dx_1 is negative and piecewise constant, discontinuous at point P.

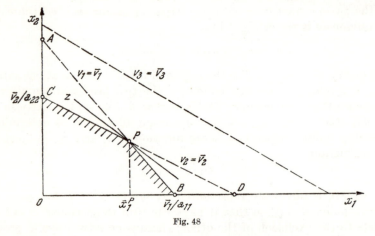

Fig. 48

The problem of maximizing total plant profit subject to the production relations (5), the capacity constraints $v_s \leq \bar{v}_s$ and the requirements $x_j \geq 0$, becomes a linear programming problem in the x_j:

$$z = \sum_{j=1}^{n} \bar{z}_j x_j = \text{maximum}$$

subject to the restrictions

$$(v_s =) \sum_{j=1}^{n} a_{sj} x_j + r_s = \bar{v}_s \qquad (s = 1, \ldots)$$

$$x_j \geq 0, \quad r_s \geq 0$$

where $\bar{z}_j = p_j - \sum_i q_i a_{ij}$ $(q_s = 0)$ is gross profit per unit of x_j and r_s is a slack variable to be interpreted as idle capacity for the s'th fixed factor. The optimum point will be one of the corners of the capacity curve. In Fig. 48 point P will represent maximum profit if the slope of the iso-profit lines $z = \bar{z}_1 x_1 + \bar{z}_2 x_2 = \text{constant}$, $dx_2/dx_1 = -\bar{z}_1/\bar{z}_2$, is greater than the slope of AB and less than that of CD; in this case, unlike the one-capacity case of Fig. 47, it pays the firm to produce both commodities because this is the only way in which the two capacities $\bar{v}_1$ and $\bar{v}_2$ can be fully utilized. It is only when $\bar{z}_1$ is large compared to $\bar{z}_2$ or vice versa that it pays to specialize (point B or C), thus leaving some of $\bar{v}_2$ or $\bar{v}_1$ idle[1].

The numerical values of the simplex coefficients of the slack variables r_s in the optimum solution can be interpreted as shadow prices associated with the fixed factors v_s. If P in Fig. 48 represents maximum profit, v_1 and v_2 will have positive shadow prices because the services of the first two capacity factors are scarce; v_3, on the other hand, is available in abundant supply so that a marginal increase in $\bar{v}_3$ would add nothing to profit, i.e., its shadow price is zero.

The shadow prices y_s can be used as an alternative instrument of optimal planning. It follows from the duality theorem of linear programming[2] that "net" unit profit—defined as gross unit profit, $\bar{z}_j$, minus the imputed cost of the amounts used of the fixed factors' services, $\sum_s y_s a_{sj}$—will be zero in those activities (processes) which are actually used in the optimum solution ($x_j > 0$), negative in those activities for which $x_j = 0$. Using this criterion for each process (product) separately, the plant optimum can be found by suboptimization; the shadow prices which are applied as accounting prices associated with the fixed factors can be determined independently—i.e., without solving the plant optimum problem above—since they are identical with the optimum solution of the dual problem of minimizing total imputed cost of the fixed factors

$$g = \sum_s \bar{v}_s y_s$$

subject to

$$\sum_s a_{sj} y_s \geq \bar{z}_j \qquad (j = 1, 2, \ldots, n)$$

for $y_s \geq 0$.

(b) Now assume that *discontinuous factor substitution* is possible in each of the processes so that each product can be made in a finite number of linear activities. In the two-product case the model for the first process becomes

$$x_1 = \sum_{j=1}^{M} \lambda_j, \quad v_{i1} = \sum_{j=1}^{M} a_{ij} \lambda_j \qquad (i = 1, 2, \ldots, m) \tag{6}$$

and for the second process

$$x_2 = \sum_{k=1}^{N} \mu_k, \quad v_{i2} = \sum_{k=1}^{N} b_{ik} \mu_k \qquad (i = 1, 2, \ldots, m) \tag{7}$$

where the λ_j and μ_k are activity levels and the a_{ij} and b_{ik} are technical coefficients characterizing the respective activities. The process models (6) and (7) are interrelated through capacity constraints of the form

---

[1] An instructive hypothetical model of this type—an automobile firm producing trucks and automobiles under four capacity restrictions—is given by DORFMAN (1953), pp. 798 ff.

[2] Cf. Appendix 1, B and Ch. III, B above.

$$v_s = v_{s1} + v_{s2} = \sum_{j=1}^{M} a_{sj} \lambda_j + \sum_{k=1}^{N} b_{sk} \mu_k \leq \bar{v}_s. \tag{8}$$

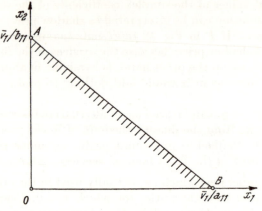

Fig. 49

The capacity curve will have the same shape as in the above case of limitationality. If there is only one capacity restriction, $v_1 \leq \bar{v}_1$, the maximum amount of x_1 that can be produced for $x_2 = 0$ is

$$x_1^{\max} = \text{Max} \, (\bar{v}_1 / a_{1j}) ;$$

assuming that the first activity has the smallest a_{1j}, we have

$$x_1^{\max} = \lambda_1 = \bar{v}_1 / a_{11},$$

cf. point B in *Fig. 49*. Similarly, assuming that $b_{11} < b_{1k}$ $(k = 2, 3, \ldots, N)$, we have at point A for $x_1 = 0$

$$x_2^{\max} = \mu_1 = \bar{v}_1 / b_{11}.$$

Then the straight line connecting A with B represents the capacity curve, $0AB$ being the region of feasible combinations of x_1 and x_2. Maximizing $x_2 = \sum_k \mu_k$ for given parametric $x_1 = \bar{x}_1$, i.e., subject to

$$\sum_{j=1}^{M} \lambda_j = \bar{x}_1$$

and subject to the capacity limitation

$$\sum_{j=1}^{M} a_{1j} \lambda_j + \sum_{k=1}^{N} b_{1k} \mu_k + r_1 = \bar{v}_1,$$

using λ_1 and μ_1 for basic variables, we get

$$x_2 = \frac{\bar{v}_1 - a_{11} \bar{x}_1}{b_{11}} - \sum_{j=2}^{M} \frac{a_{1j} - a_{11}}{b_{11}} \lambda_j - \sum_{k=2}^{N} \frac{b_{1k} - b_{11}}{b_{11}} \mu_k - \frac{1}{b_{11}} r_1$$

where the simplex coefficients are all negative so that x_2 is a maximum for $\lambda_j = \mu_k = 0$ $(j = 2, 3, \ldots, M; \ k = 2, 3, \ldots, N)$, $r_1 = 0$; the solution is feasible for $\bar{x}_1 \leq \bar{v}_1/a_{11}$. For parametric $\bar{x}_1$ the solution represents the line segment AB, along which $dx_2/d\bar{x}_1 = -a_{11}/b_{11}$.

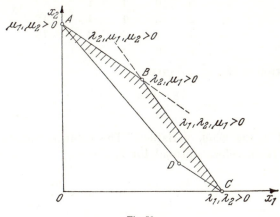

Fig. 50

The capacity curve will be kinked if there is more than one capacity limit. For $\bar{x}_1 = 0$, i.e., all $\lambda_j = 0$, the maximum of $x_2 = \sum_k \mu_k$ subject to (8) for $s = 1, 2$ will normally be some combination of two activities—say, μ_1 and μ_2—since there are two linear restrictions, cf. point A in *Fig. 50*. For positive $\bar{x}_1 = \sum_j \lambda_j$ there will be three side conditions so that a third activity representing the parameter $\bar{x}_1$ is required in the optimal basis, say, λ_2. For increasing $\bar{x}_1$, λ_2 increases at the expense of μ_1 or μ_2 (or both)—cf. AB in Fig. 50—until one of these variables, say, μ_2, becomes zero. Beyond this point (B) the basis is no longer feasible and μ_2 will have to be replaced by, say, λ_1 in the basis; for increasing $\bar{x}_1$ the new basic optimal solution $(\lambda_1, \lambda_2, \mu_1)$ moves from B to C where C is the point of maximum x_1 for x_2 $(= \mu_1) = 0$. $dx_2/d\bar{x}_1$ is negative and piecewise constant. (With three capacity restrictions there will be two kinks—where $dx_2/d\bar{x}_1$ is discontinuous—on the capacity curve since the transition from a parametric optimal basic of three μ_k to one of three λ_j requires two shifts in the basis, for example,

$$(\mu_1, \mu_2, \mu_3)$$
$$\mu_1, \mu_2, \mu_3, \lambda_1$$
$$(\mu_1, \mu_2, \lambda_1)$$
$$\mu_1, \mu_2, \lambda_1, \lambda_2$$
$$(\mu_1, \lambda_1, \lambda_2)$$
$$\mu_1, \lambda_1, \lambda_2, \lambda_3$$
$$(\lambda_1, \lambda_2, \lambda_3),$$

and so on for any number of capacities.)

For example, if the matrix of technical coefficients is as follows:

	λ_1	λ_2	μ_1	μ_2	
x_1	1	1	0	0	
x_2	0	0	1	1	
v_1	2	3	1	4	≤ 12
v_2	4	1	4	2	≤ 18
v_3	1	6	8	2	

the basic solution in $(\lambda_2, \mu_1, \mu_2)$ gives

$$x_2 = \mu_1 + \mu_2 = \frac{78 - 9\bar{x}_1}{14} - \frac{1}{2}\lambda_1 - \frac{1}{7}r_1 - \frac{3}{14}r_2$$

(where r_1 and r_2 are slack variables). The solution satisfies the simplex criterion and is therefore optimal for $\lambda_1 = r_1 = r_2 = 0$ provided that it is feasible,

$$\lambda_2 = \bar{x}_1 \geq 0, \quad \mu_1 = \frac{48 + 2\bar{x}_1}{14} \geq 0, \quad \mu_2 = \frac{30 - 11\bar{x}_1}{14} \geq 0,$$

i.e., for $0 \leq \bar{x}_1 \leq 30/11$ (cf. AB in Fig. 50). For $\bar{x}_1 = 0$ we have $x_2 = 78/14$, the absolute maximum of x_2 (point A). For $x_1 \geq 30/11$ the solution in $(\lambda_1, \lambda_2, \mu_1)$ gives

$$x_2 = \mu_1 = \frac{54 - 10\bar{x}_1}{7} - 2\mu_2 - \frac{3}{7}r_1 - \frac{1}{7}r_2$$

which is optimal for $\mu_2 = r_1 = r_2 = 0$ if

$$\lambda_1 = \frac{11\bar{x}_1 - 30}{7} \geq 0, \quad \lambda_2 = \frac{30 - 4\bar{x}_1}{7} \geq 0, \quad \mu_1 = \frac{54 - 10\bar{x}_1}{7} \geq 0,$$

that is, for

$$30/11 \leq \bar{x}_1 \leq 54/10$$

(BC in Fig. 50) where $x_1 = 54/10$ (point C) is the maximum amount that can be produced for $x_2 = 0$. The slope of the capacity curve, $dx_2/d\bar{x}_1$, shifts at B from $-9/14$ to $-10/7$.

Cases of this type[1] differ from that of Fig. 48 in that both capacities are fully utilized ($v_1 = \bar{v}_1$, $v_2 = \bar{v}_2$, i.e., $r_1 = r_2 = 0$) everywhere on the capacity curve, and even in part of the interior of the feasible region. For example, in Fig. 50 (which corresponds to the numerical example) both capacities are exhausted at the interior point D where $x_1 = \lambda_1 = 4$, $x_2 = \mu_2 = 1$. Any point which can be expressed as a convex combination of two or more points satisfying $v_1 = \bar{v}_1$, $v_2 = \bar{v}_2$ with two positive activity levels will also exhaust the capacities, so that the convex polygon $ABCD$ represents the region of full capacity utilization.

---

[1] An empirical example of multiple production where each product is manufactured in a finite number of activities and the products are interdependent through the joint use of capacities is given by CHARNES, COOPER and FARR (1953), cf. also Ch. III, B and IX, B above.

In the case of two fixed factors the capacity curve can also be derived from a box diagram, being defined along the locus of points at which two isoquants $x_1 = \bar{x}_1$ and $x_2 = \bar{x}_2$ are "tangent" to each other. Such a contract curve is shown as the broken line $A\,B\,C$ in *Fig. 51*, where A and C are origins for isoquant maps of x_1 and x_2 respectively; the sides of the box (AQ and $A\,P$) represent $\bar{v}_1$ and $\bar{v}_2$.

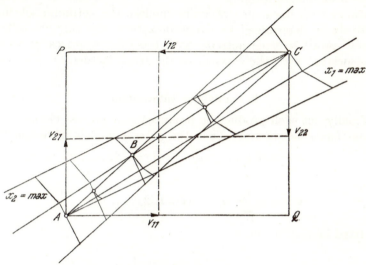

Fig. 51

The point of maximal profit, as determined by maximizing

$$z = \sum_{j=1}^{M}\left(p_1 - \sum_{i=1}^{m} q_i a_{ij}\right)\lambda_j + \sum_{k=1}^{N}\left(p_2 - \sum_{i=1}^{m} q_i b_{ik}\right)\mu_k$$

subject to the capacity restrictions (8), does not have to be a point on the capacity curve as it did in the limitational case[1]. The number of activities—structural or slack activities—used in the optimal allocation is equal to, or in special cases less than, the number of fixed factors; whether it pays to specialize—producing one output in several activities—or to produce several products jointly, and whether the optimum point is on the boundary or in the interior of the feasible region, depends on the prices p_j and q_i and on the coefficients a_{ij} and b_{ik} of the current inputs[2]. In the numerical example above, the prices $p_1 = 12$, $p_2 = 15$, $q_3 = 1$ (the prices of the fixed factors being zero) lead to the profit function

---

[1] For one thing, the geometrical method of finding the optimum illustrated in Fig. 48 does not apply since profit z cannot be expressed as a linear function of x_1 and x_2; unit profit can be calculated for each *activity* but not for each *product*, cf. the profit expression above.

[2] For example, if the coefficients a_{i1} and b_{i2} for the variable inputs $v_3, \ldots, v_m$ are sufficiently small (as compared with those in the other activities), the best solution may be to use activities λ_1 and μ_2 in the numerical two-capacity example above, resulting in an interior point of the region characterized by $v_1 = \bar{v}_1$, $v_2 = \bar{v}_2$ (cf. above).—It is not even certain that both capacities will be fully utilized.

$$z = 11\lambda_1 + 6\lambda_2 + 7\mu_1 + 13\mu_2$$

which is a maximum for $\lambda_1 = 4$, $\mu_2 = 1$—i.e., at point D in Fig. 50—since the profit function in terms of the non-basic variables,

$$z = 57 - 3\lambda_2 - \frac{3}{2}\mu_1 - \frac{5}{2}r_1 - \frac{3}{2}r_2,$$

satisfies the simplex criterion.

As always with linear programming models, the optimum solution may alternatively be determined by suboptimization, applying the criterion of zero "net" profit to each activity with the shadow prices as accounting prices associated with the services of the fixed factors.

3. Continuous Substitution

(a) Finally, let each product be produced in a process characterized by *continuous factor substitution*. Then the process models have the form

$$x_j = f_j(v_{1j}, \ldots, v_{mj}) \qquad (j = 1, 2, \ldots, n) \tag{9}$$

where

$$v_i = \sum_{j=1}^{n} v_{ij} \qquad (i = 1, 2, \ldots, m); \tag{10}$$

for the fixed factors we have

$$v_s = \sum_{j=1}^{n} v_{sj} \leq \bar{v}_s.\,^{[1]} $$

(9)–(10) is a parametric plant production model. The number of parameters $v_{ij}(\geq 0)$ is generally greater than the number of equations ($m \cdot n > m + n$ for $m, n > 2$) so that the v_{ij} cannot be eliminated to give a plant model in the x_j and v_i only.

(b) Fig. 45 may be taken as illustrating the special two-product case

$$x_1 = f(v_{11}, v_2), \quad x_2 = g(v_{12}, v_3) \tag{11}$$

with a single capacity restriction

$$v_1 = v_{11} + v_{12} \leq \bar{v}_1\,^{[2]}. \tag{12}$$

The capacity curve, as determined by maximizing x_2 for given parametric $\bar{x}_1$ and subject to $v_1 \leq \bar{v}_1$, will be a straight line—as in the cases of Fig. 47 and 49—if the functions f and g are homogeneous of degree one. This is shown in *Fig. 52*, where A and B are the origins corresponding to the isoquant maps, AB being equal to the capacity limit $\bar{v}_1$. The maximum amount of x_2 for given $\bar{x}_1$ is seen to result from expanding $\bar{x}_1$ along the path AP—the boundary of the region of substitution — and decreasing x_2 along QB, where x_2 is proportional to v_{12} and $\bar{x}_1$ is proportional to $v_{11} = \bar{v}_1 - v_{12}$ so that x_2 varies linearly with $\bar{x}_1$.

---

[1] Cf. Bordin (1944), p. 90, Ricossa (1954), and Pfouts (1961).

[2] This special case can be reduced to a single-equation plant model

$$v_1 = \varphi(x_1, v_2) + \psi(x_2, v_3) \quad (\leq \bar{v}_1)$$

where the functions φ and ψ are derived from (11) by solving for v_{11} and v_{12} respectively.

The same result can be derived analytically be maximizing the Lagrangian

$$L = g(v_{12}, v_3) + u_1 \cdot (\bar{x}_1 - f(v_{11}, v_2)) + u_2 \cdot (\bar{v}_1 - v_{11} - v_{12}).$$

Necessary conditions are

$$g_1' - u_2 = 0, \quad g_3' = 0, \quad -u_1 f_1' - u_2 = 0, \quad f_2' = 0 \qquad (13)$$

where the second and the fourth equation uniquely determine the relative factor proportions, v_{12}/v_3 and v_{11}/v_2 (cf. BQ and AP in Fig. 52). The other two equations determine the multipliers $u_2 = g_1'$ and $u_1 = -g_1'/f_1'$. The

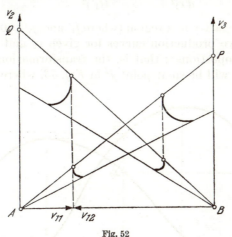

Fig. 52

latter is to be interpreted as the (constant) slope of the capacity curve. By the homogeneity of the functions (Euler's theorem) we have for g_3' and $f_2' = 0$

$$\bar{x}_1 = v_{11} f_1', \quad x_2 = v_{12} g_1' = (\bar{v}_1 - v_{11}) \cdot g_1'$$

which leads to

$$x_2 = \bar{v}_1 g_1' - (g_1'/f_1') \cdot \bar{x}_1$$

where f_1' and g_1' are constant under the variations considered so that x_2 is a linear function of $\bar{x}_1$, the slope being

$$dx_2/d\bar{x}_1 = -g_1'/f_1' = u_1 < 0.\,[1]$$

For any given set of input amounts $v_i = \bar{\bar{v}}_i$ $(i = 1, 2, 3)$, $\bar{\bar{v}}_1 \leq \bar{v}_1$, model (11)–(12) defines a *transformation curve* as the locus of feasible combinations of x_1 and x_2. Product substitution along the curve—analogous to factor substitution along an isoquant for given output—is made possible by reallocation of $\bar{\bar{v}}_1$ between the two processes. Differentiating (11)–(12) for $v_i = \bar{\bar{v}}_i$ we have

---

[1] The solution given by (13) with (11) for $v_1 = \bar{v}_1$ is seen to satisfy the Kuhn-Tucker conditions for maximum of x_2 subject to (11) for given $\bar{x}_1$ and to the *inequality* $v_1 \leq \bar{v}_1$ (cf. Appendix 1, C): the multiplier u_2 associated with the inequality is positive $(= g_1' > 0)$ and no sign restrictions are put on u_1 (which is negative here) since the corresponding side condition is an equality.

$$dx_1 = f_1' \, dv_{11}, \quad dx_2 = g_1' \, dv_{12}, \quad dv_{11} + dv_{12} = 0,$$

i.e.,

$$\frac{dx_2}{dx_1} = -\frac{g_1'}{f_1'},$$

where $-dx_2/dx_1 = g_1'/f_1'$ is the marginal rate of product substitution at the point. At a point where f_1' and g_1' are both positive, dx_2/dx_1 is negative. The second derivative,

$$\frac{d^2x_2}{dx_1^2} = \frac{g_{11}'' + f_{11}'' \cdot (g_1'/f_1')}{(f_1')^2},$$

will normally be negative in a region (where f_1' and $g_1' > 0$, f_{11}'' and $g_{11}'' < 0$) when the respective production curves for given v_2 and v_3 conform to the law of variable proportions[1]; that is, the transformation curve is concave to the origin. This will be so at point P in *Fig. 53*, where AB represents $\bar{\bar{v}}_1$.

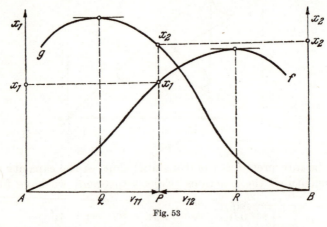

Fig. 53

However, for sufficiently large $\bar{\bar{v}}_1$ there may be inefficient points where f_1' and g_1' have opposite signs so that $dx_2/dx_1 > 0$; this will be so in the intervals AQ and RB in Fig. 53. On the other hand, for sufficiently small

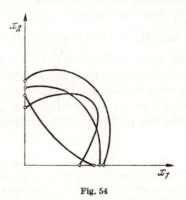

Fig. 54

---

[1] Cf. Ch. IV, B and Appendix 2.

$\bar{\bar{v}}_1$ ($\bar{\bar{v}}_2$ and $\bar{\bar{v}}_3$ still the same) the interval AB will be so short that f_{11}'' and g_{11}'' (as well as f_1' and g_1') are positive everywhere, i.e., the transformation curve is convex to the origin. A family of transformation curves for different values of $\bar{\bar{v}}_1$ ($\bar{\bar{v}}_2$ and $\bar{\bar{v}}_3$ being the same) is shown in *Fig. 54*.

The capacity curve is the envelope of the transformation curves for which $\bar{\bar{v}}_1$ is equal to the capacity limit, $\bar{v}_1$. Any point on the capacity line is also a point on some transformation curve for $\bar{\bar{v}}_1 = \bar{v}_1$ and the former is tangent to the latter at the point since $dx_2/dx_1 = -g_1'/f_1'$ for both curves, as is illustrated (with a linear capacity curve) in *Fig. 55*.

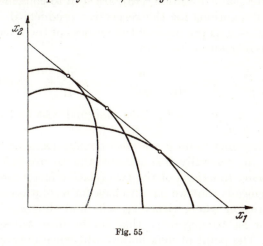

Fig. 55

(c) Maximizing total profit $z = p_1 x_1 + p_2 x_2 - q_2 v_2 - q_3 v_3$ subject to (11) we get the necessary conditions

$$f_1' = 0, \quad p_1 f_2' - q_2 = 0$$
$$g_1' = 0, \quad p_2 g_3' - q_3 = 0, \tag{14}$$

four equations in the four input variables. If a solution exists and is feasible (i.e., if $v_{11}, v_{12} \geq 0$, $v_{11} + v_{12} \leq \bar{v}_1$), it represents the optimum allocation of the fixed factor. Since f_2' and g_3' are positive for $q_2, q_3 > 0$, the solution cannot be a point on the capacity curve, cf. (13). The shadow price of the fixed factor is zero if $v_1 < \bar{v}_1$ since a marginal increase in capacity would add nothing to profit.

If the solution (14) exists but violates the capacity restriction (i.e., if $v_{11} + v_{12} > \bar{v}_1$), a new solution will have to be determined by maximizing z subject to $v_1 = \bar{v}_1$. Maximizing the Lagrangian $L = z + u \cdot (\bar{v}_1 - v_{11} - v_{12})$ we get a solution determined by the conditions

$$p_1 f_1' - u = 0, \quad p_1 f_2' - q_2 = 0$$
$$p_2 g_1' - u = 0, \quad p_2 g_3' - q_3 = 0 \tag{15}$$

and

$$v_{11} + v_{12} = \bar{v}_1. \tag{16}$$

The solution to (15)–(16), provided that it exists, is optimal if it satisfies the non-negativity requirements $v_{11} \geq 0$, $v_{12} \geq 0$. The Lagrange multiplier u

$(= p_1 f_1' = p_2 g_1'$, the common value of the fixed factor's marginal productivity in the two product directions) can be identified with the shadow price, y, since for variations satisfying (15)–(16) we have

$$dz = p_1 f_1' \, dv_{11} + p_2 g_1' \, dv_{12} + (p_1 f_2' - q_2) \, dv_2 + (p_2 g_3' - q_3) \, dv_3$$
$$= u \cdot (dv_{11} + dv_{12}) = u \cdot d\bar{v}_1$$

or
$$y = dz / d\bar{v}_1 = u.$$

The conditions for plant optimum, (15), are seen to coincide with the combined suboptimal solutions for the respective products if and only if the shadow price y is used as the price of the services of the fixed factor. Conditions (15) can be rearranged to give

$$p_1 = \frac{y}{f_1'} = \frac{q_2}{f_2'}, \quad p_2 = \frac{y}{g_1'} = \frac{q_3}{g_3'}, \quad p_1 = p_2 \frac{g_1'}{f_1'} = -p_2 (dx_2 / dx_1);$$

for each product, the price is equal to marginal cost and to the marginal opportunity cost of reallocating the fixed factor.

Now if the functions f and g are both homogeneous of the first degree, conditions (14) are generally inconsistent and so are (15), the marginal productivities being functions of the two relative factor proportions v_{11}/v_2 and v_{12}/v_3. This means that we have to look for a solution on the boundary of the feasible region where either $v_{11} = 0$ (and thus $v_2 = x_1 = 0$, $v_{12} = \bar{v}_1$, $v_3 > 0$, $x_2 > 0$) or $v_{12} = 0$ ($v_{11} = \bar{v}_1$), that is, the firm will specialize in one of the products. The point of maximum profit when only x_1 is produced is determined by

$$p_1 f_2' - q_2 = 0 \quad (v_{11} = \bar{v}_1); \tag{17}$$

the condition for maximum profit in the other direction is

$$p_2 g_3' - q_3 = 0 \quad (v_{12} = \bar{v}_1). \tag{18}$$

If profit z_1 corresponding to (17) is greater than z_2 as determined by (18), the shadow price will be

$$y = p_1 f_1'$$

and total "net" profit, using y for the price of the fixed factor, will be zero $(z_1 - \bar{v}_1 p_1 f_1' = 0)$ [1]; with the same y, net profit would have been negative if x_2 had been produced instead (point (18)). Whether x_1 or x_2 will actually be produced will depend on the prices.

(d) More general cases may be treated on similar lines. For example, a model of the form

$$x_1 = f(v_{11}, v_{21}, v_3), \quad x_2 = g(v_{12}, v_{22}, v_4) \tag{19}$$

with two capacity restrictions

$$v_1 = v_{11} + v_{12} \leq \bar{v}_1, \quad v_2 = v_{21} + v_{22} \leq \bar{v}_2 \tag{20}$$

leads to a capacity curve determined by the conditions

---
[1] Cf. Ch. IV, F.

$$\frac{f_2'}{f_1'}=\frac{g_2'}{g_1'}, \quad f_3'=0, \quad g_4'=0$$

where $dx_2/d\bar{x}_1=-g_1'/f_1'\,(=-g_2'/f_2')$ at a point of the curve.

Optimization problems subject to two capacities can be treated in a way similar to that used above, turning the inequality constraints into equations ($v_s=\bar{v}_s$, $v_{sj}=0$) until a feasible solution is found which satisfies the necessary conditions for maximum[1].

C. Multi-Product Processes

1. Joint Production and Cost Allocation

Let us now consider the case of truly joint production where the products are made in a single process—such as P in Fig. 46—which is not additively separable[2]. The total amount of each input used, and so also total cost, cannot be allocated to the respective outputs[3]—the various methods of allocation used in practice for pure accounting purposes are all arbitrary, having no technological foundation—so that the production model will have to be a plant model, expressing the technological relationships between plant inputs and outputs, v_i and x_j [4]. The cost function will generally be a function of all outputs, $c=c(x_1, x_2, \ldots, x_n)$, defined along the locus of least-cost points for parametric x_j and subject to the production relations.

In the following we shall deal with some of the more important types of joint-production models, ranging from the extreme case of fixed output proportions—analogous to limitational inputs—to the opposite extreme, a single-equation model representing continuous product and factor substitutability.

2. Fixed Output Proportions

Let the process be such that the products are perfectly complementary, being made in technologically fixed proportions,

$$x_j=b_j \cdot x_1 \qquad (j=2, 3, \ldots, n) \tag{21}$$

where the level of output as expressed by, say, x_1 depends upon the inputs so that the plant production model consists of (21) and a single-product

---

[1] More generally, problems of this kind can be treated by use of the Kuhn-Tucker saddle point conditions. For a general treatment on these lines of cost minimization problems subject to models of the type (9)–(10) and to capacity constraints, see PFOUTS (1961).

[2] Cf. WALTERS (1963), p. 43.

[3] However, when the joint outputs of a process are further processed in separate processes (representing, for example, finishing operations), the inputs of the latter processes are specific to the respective products; cf. WINDING PEDERSEN (1949), pp. 92 f.

The limiting case where the joint products of the first process are identical—e.g. a crude product to be differentiated into several distinct grades by further separate processing—is an example of alternative rather than joint production.

[4] A wide variety of cases of joint production are described by RIEBEL (1955), though the precise nature of the respective models is not indicated.

model in x_1 and the v_i. The latter model may be characterized by continuous substitution, or the v_i may be proportional to x_1 so that the model
consists of a single activity; other models are also possible[1]. In any case, the
transformation curve for any given (feasible) factor combination will be a
point in product space, cf. point P in *Fig. 56*. It is possible to get any combination on AP or PB (or in the interior of the rectangle $0APB$) by producing the combination P and allowing some of x_1 or x_2 to go to waste (or
destroying it)[2]. However, since the output variables x_1 and x_2 are defined
as quantities actually produced, APB is not the transformation curve
belonging to (21); only point P belongs to the production model[3].

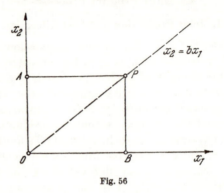

Fig. 56

Whatever kind of model connects x_1 with the v_i, the multi-product model
of the plant can be treated as a single-output case in these $m+1$ variables,
only the price associated with x_1 (revenue per unit of x_1) becomes

$$\frac{\sum\limits_{j=1}^{n} p_j x_j}{x_1} = p_1 + \sum\limits_{j=2}^{n} b_j p_j.$$

3. Linear Models

Now if there is more than one linear activity which produces the n products jointly, the model will be of the form

$$x_j = \sum_{k=1}^{N} b_{jk} \lambda_k \qquad (j=1, 2, \ldots, n)$$

$$v_i = \sum_{k=1}^{N} a_{ik} \lambda_k \qquad (i=1, 2, \ldots, m)$$

(22)

where the relative output proportions x_j/x_1 $(j=2, 3, \ldots, n)$, as well as the
relative factor combination, can be varied within limits by varying the ratios

---

[1] Cf. Chs. III–VI.

[2] Cf. SCHMIDT (1939), pp. 288 f.

[3] Cf. the analogous problem whether the isoquants are L-shaped or single points in cases
of limitational inputs (Ch. III, A).

of the (non-negative) activity levels λ_k ($k = 1, 2, \ldots, N$) as illustrated by the two-product, four-activity case of *Fig. 57*. The feasible product region is bounded by the two extreme activity half-lines and by the capacity curve, which will be a broken line. In Fig. 57, $v_1 = \bar{v}_1$ at A, B, C, and D and on the segments connecting them, whereas the second capacity factor is fully utilized at E, F, G, and H; the capacity curve—the locus of maximum x_2 for given parametric $\bar{x}_1$ and for $v_1 \leq \bar{v}_1$, $v_2 \leq \bar{v}_2$—will be $APFQH$.[1]

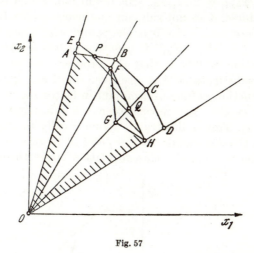

Fig. 57

The point where total profit z—a linear function of the activity levels λ_k—is a maximum subject to two capacity restrictions will generally be characterized by two positive activity levels (not necessarily both structural)[2]. It does not have to be a point on the capacity curve; for example, the activity through G may be used in the optimal solution if it has sufficiently small input coefficients for the variable inputs.

4. Continuous Product and Factor Substitution

(a) The "classical" model of joint production is the production function

$$F(x_1, x_2, \ldots, x_n, v_1, v_2, \ldots, v_m) = 0, \tag{23}$$

an obvious generalization of the single-product model $x = x(v_1, v_2, \ldots, v_m)$[3].

---

[1] Point G is not on the capacity curve since both x_1 and x_2 are greater at Q, which represents a convex combination of F and H.—Note, however, that G cannot be dismissed as inefficient: Q is "better" than G with respect to the factor v_2, but it is quite possible that it uses more of other (variable) inputs not specified in the figure. The concept of inefficiency applies to transformation curves, not to capacity curves.

[2] Even if only one structural activity level is positive in the optimum solution, both products will be made (except in the special case where one of the output coefficients in the activity is zero so that the activity half-line coincides with one of the axes). The *activities* may be said to be alternative in the above sense but the *products* are not.

[3] For modern treatments see, for example, CARLSON (1939), Ch. 5; HICKS (1946), Ch. 6 with Appendix; DORFMAN (1951), Ch. 1; and ALLEN (1956), Ch. 18, 2.

Alternatively the function may be written in the less symmetrical form

$$x_n = f(x_1, x_2, \ldots, x_{n-1}, v_1, v_2, \ldots, v_m). \tag{23a}$$

The function F, or f, assumed to possess continuous first- and second-order partial derivatives in the non-negative region, is defined as giving the maximum amount of any product—say, x_n—that can be produced with the given technology for any feasible combination of $m + n - 1$ independent variables $x_1, \ldots, x_{n-1}, v_1, \ldots, v_m$. This definition, which is adopted to ensure single-valuedness, does not rule out inefficient points from the production function; there may exist feasible points where some marginal productivity $\delta x_j / \delta v_i \ (= -F_{vi}' / F_{xj}')$ is negative or where some $\delta x_j / \delta x_k > 0$ [1].

The interpretation of (23) is that the products, as well as the factors, are continuously substitutable. For any given set of $x_j = \bar{\bar{x}}_j \ (j = 1, 2, \ldots, n)$, the production function defines an isoquant in factor space; for given $v_i = \bar{\bar{v}}_i$ $(i = 1, 2, \ldots, m)$ we get a transformation curve indicating the feasible combinations of the x_j. This is illustrated geometrically in *Fig. 58*, where the *isoquant* through Q corresponds to the *point* P in product space and the factor combination Q gives the transformation curve through P [2].

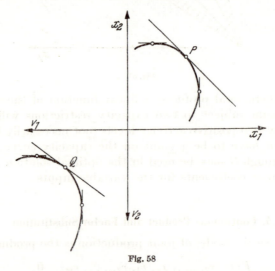

Fig. 58

The efficient region, or the region of (product and factor) substitution, is defined by

$$\delta x_j / \delta v_i = -F_{vi}' / F_{xj}' \geq 0$$
$$\delta x_j / \delta x_k = -F_{xk}' / F_{xj}' \leq 0$$
$$\delta v_i / \delta v_h = -F_{vh}' / F_{vi}' \leq 0$$

for all i, j, h, k; that is, the region where the F_{vi}' are all ≥ 0 and the F_{xj}' all ≤ 0 (or vice versa [3]). The isoquants and the transformation curves, which

---

[1] Cf. Ch. II and IV, A above.

[2] The assumption of single-valuedness no longer precludes the intersection of isoquants (or transformation curves) as it did in the single-product case.

[3] All points satisfying (23) also satisfy the equation $-F = 0$.

are negatively inclined in the interior of the region, will be assumed to be respectively convex and concave to the origin as shown in Fig. 58 for the two-product, two-factor case[1].

The case of constant returns to scale implies that the function f is homogeneous of the first degree, or that F is homogeneous of degree zero, so that the partial derivatives F_{xj}', F_{vi}' are homogeneous of degree minus one. Then we can write

$$F = \varphi\left(\frac{x_1}{v_1}, \ldots, \frac{x_n}{v_1}, \frac{v_2}{v_1}, \ldots, \frac{v_m}{v_1}\right) = 0 \qquad (24)$$

where φ is some function, and

$$F_{xj}' = \frac{1}{v_1}\delta\varphi/\delta(x_j/v_1) \qquad (j = 1, 2, \ldots, n)$$

$$F_{vi}' = \frac{1}{v_1}\delta\varphi/\delta(v_i/v_1) \qquad (i = 2, 3, \ldots, m) \qquad (25)$$

$$F_{v1}' = \frac{-1}{v_1}\cdot\left(\sum_{j=1}^{n}\frac{x_j}{v_1}\delta\varphi/\delta(x_j/v_1) + \sum_{i=2}^{m}\frac{v_i}{v_1}\delta\varphi/\delta(v_i/v_1)\right).[2]$$

This case can be regarded as a limiting case of a linear model where the number of activities is infinite. The opposite extreme case is that of fixed output proportions and limitational inputs with fixed coefficients, i. e., a single multi-product activity.

The capacity curve is defined by maximizing, say, x_n for given x_1, x_2, $\ldots$, x_{n-1} and subject to the capacity restrictions $v_s \leq \bar{v}_s$. In the two-factor, two-product case with a single capacity limit $v_1 \leq \bar{v}_1$ the capacity curve is determined by the production function

$$x_2 = f(x_1, v_1, v_2)$$

with the condition

$$f_{v2}' = 0$$

for $v_1 = \bar{v}_1$ and for given parametric $x_1 = \bar{x}_1$. Since

$$dx_2 = f_{x1}'\,d\bar{x}_1 + f_{v2}'\,dv_2 = f_{x1}'\,d\bar{x}_1$$

along the curve, the slope of the curve is

$$dx_2/d\bar{x}_1 = f_{x1}'(\bar{x}_1, \bar{v}_1, v_2)$$

so that the capacity curve is the envelope of the family of transformation curves for $v_1 = \bar{v}_1$ as shown in *Fig. 59*.

(b) To find necessary conditions for maximum profit, maximize the Lagrangian $L = z + u \cdot F$; this gives[3]

---

[1] The definition of convexity and concavity in the general case is obvious from that given for the single-product case in Ch. IV, A.

[2] The expression for F_{v1}' has been derived by use of Euler's theorem,

$$v_1 F_{v1}' + \sum_{i=2}^{m} v_i F_{vi}' + \sum_{j=1}^{n} x_j F_{xj}' = 0.$$

[3] Cf. ALLEN (1956), pp. 613 ff., or HICKS (1946), pp. 86 ff. and 319 f.

$$p_j + u \cdot F_{xj}' = 0 \qquad (j = 1, 2, \ldots, n) \tag{26}$$

$$-q_i + u \cdot F_{vi}' = 0 \qquad (i = 1, 2, \ldots, m) \tag{27}$$

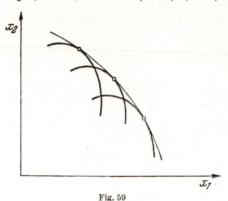

Fig. 59

(where the prices of the capacity factors are zero). If a solution to (23) and (26)–(27) exists and is feasible—i.e., if it is non-negative and respects the capacity constraints $v_s \leq \bar{v}_s$—it represents the point of optimum allocation[1]. (26)–(27) can be rearranged to give

$$p_j = \frac{-q_i}{F_{vi}' / F_{xj}'} = \frac{q_i}{\delta x_j / \delta v_i} \qquad (i = 1, 2, \ldots, m; \ j = 1, 2, \ldots, n), \tag{28}$$

that is, for each particular product the price is equal to the partial marginal cost in every factor direction; it is also equal to marginal opportunity cost in every product direction,

$$p_j = -p_k \frac{\delta x_k}{\delta x_j} \qquad (j, k = 1, 2, \ldots, n). \tag{29}$$

At the optimum point the slope of the tangent to the transformation curve (the marginal rate of product substitution) is equal to the price ratio,

$$p_j / p_k = -\delta x_k / \delta x_j \qquad (j, k = 1, 2, \ldots, n),$$

and similarly for the marginal rate of substitution on the isoquant we have

$$q_i / q_h = -\delta v_h / \delta v_i \qquad (i, h = 1, 2, \ldots, m).$$

The same results could have been obtained by first minimizing cost for given parametric x_j, thus determining the joint cost function $c = c(x_1, \ldots, x_n)$, and next maximizing total revenue minus cost. Minimizing the Lagrangian $L = \sum_i q_i v_i - u \cdot F$ for given $x_j = \bar{x}_j$ we get (27) above:

$$q_i - u \cdot F_{vi}' = 0 \, ;$$

since at a point on the expansion path satisfying these conditions and the production function we have

---

[1] Sufficient second-order conditions are given by ALLEN and HICKS, *loc. cit.* These stability conditions are automatically satisfied and ensure a unique maximum if the isoquants are convex and the transformation curves are concave.

$$dc = \sum_i q_i \, dv_i = u \cdot \sum_i F_{vi}' \, dv_i = -u \cdot \sum_j F_{xj}' \, d\bar{x}_j$$

it follows that $-u F_{xj}'$ is the partial marginal cost of product no. j along the expansion path,

$$c_j' = \delta c / \delta x_j = -u F_{xj}'.$$

At the point of maximum profit we must have

$$p_j - c_j' = 0 \qquad (j = 1, 2, \ldots, n)$$

which leads to conditions (26)[1].

Now if the production function F is homogeneous of degree zero, there exists no solution to conditions (26)–(27) and the production function (23) since there are $m + n + 1$ equations in $m + n$ variables, u / v_1 and the $m + n - 1$ relative proportions x_j / v_1, v_i / v_1 $(i = 2, 3, \ldots, m)$[2]. In order to get a solution the capacity restrictions will have to be taken explicitly into account. With a single capacity limit, maximize the Lagrangian $L = z + u \cdot F$ for $v_1 = \bar{v}_1$. This leads to conditions similar to (26)–(27), except that we have got rid of the first equation of (27) so that the system is no longer overdetermined. With this modification the interpretation of the optimum solution is similar to that given above[3].

The solution implicitly determines the shadow price of the fixed factor, y. For arbitrary variations in the x_j and v_i we have

$$dz = \sum_{j=1}^{n} p_j \, dx_j - \sum_{i=2}^{m} q_i \, dv_i;$$

if the variations are to satisfy the production function (23) in differential form for $v_1 = \bar{v}_1$,

$$\sum_{j=1}^{n} F_{xj}' \, dx_j + F_{v1}' \, d\bar{v}_1 + \sum_{i=2}^{m} F_{vi}' \, dv_i = 0$$

(where $d\bar{v}_1$ is a marginal increase in the fixed factor's capacity) as well as the optimum conditions (26)–(27) for $i = 2, 3, \ldots, m$, this reduces to

$$dz = u \cdot F_{v1}' \cdot d\bar{v}_1$$

so that

$$y = u \cdot F_{v1}'.$$

---

[1] STACKELBERG (1932), p. 63, who does not specify the underlying production function, defines an expansion path in product space as the locus of tangency between cost and revenue indifference curves.

[2] The situation is quite analogous to that of profit maximization under a linear-homogeneous single-output production function, cf. Ch. IV, E–F.

[3] For positive factor prices the optimum solution is *not* a point on the capacity curve; by (28) the marginal productivities of the variable inputs are positive $(\delta x_m / \delta v_i = q_i / p_m > 0)$ whereas $\delta x_m / \delta v_i = 0$ on the capacity curve (taking x_m as the dependent variable). The optimum solution is a point on a transformation curve where $v_1 = \bar{v}_1$, but it is not the point where the curve touches the envelope. This is analogous to profit maximization subject to a single-output production function and to a capacity restriction $v_1 \leq \bar{v}_1$, cf. Ch. IV, F above. A point where the capacity of the fixed factor is fully utilized $(v_1 = \bar{v}_1)$ does not necessarily represent maximum output, $x = x^{\max}$.

Since, by (26), $u = -p_j/F_{xj}'$ $(j = 1, 2, \ldots, n)$, it follows that

$$y = p_j \cdot (\delta x_j / \delta \bar{v}_1),$$

that is, the shadow price—the marginal contribution to profit of an increase in $\bar{v}_1$—is equal to the value of the marginal productivity of v_1 in every product direction. Alternatively, by (27) we get

$$y = -q_i \cdot (\delta v_i / \delta \bar{v}_1) \qquad (i = 2, 3, \ldots, m)$$

so that the tangency conditions for the isoquant are satisfied for *all* pairs of inputs if the shadow price is applied to the services of the fixed factor. Using this price, the marginal conditions (27) and (28) are satisfied also for $i = 1$; any other price will lead to an inoptimal allocation.

By the homogeneity of the function F (Euler's theorem) we have

$$\sum_{j=1}^{n} x_j F_{xj}' + \bar{v}_1 F_{v1}' + \sum_{i=2}^{m} v_i F_{vi}' = 0,$$

which with conditions (26)–(27) $(i = 2, 3, \ldots, m; j = 1, 2, \ldots, n)$ leads to

$$\sum_{j=1}^{n} p_j x_j - u \cdot \bar{v} \cdot F_{v1}' - \sum_{i=2}^{m} q_i v_i = 0$$

where

$$u \bar{v}_1 F_{v1}' = y \bar{v}_1$$

is the total imputed value of the fixed factor's services; using the shadow price, total "net" profit is zero[1].

5. Shadow Products

Finally, let us assume that there is a separate production function for each output in terms of the same inputs,

$$x_j = f_j(v_1, v_2, \ldots, v_m) \qquad (j = 1, 2, \ldots, n).[2] \qquad (30)$$

A model of this kind[3] implies that the output proportions x_j/x_1 $(j = 2, 3, \ldots, n)$ can be varied only through variations in the factor combination: the transformation curve for given $v_i = \bar{\bar{v}}_i$ is a single point in product space but the position of the point depends on the parameters $\bar{\bar{v}}_i$. Conversely, for given values of the x_j there are n equations in the v_i so that the model is characterized by constrained factor substitution (for $n < m$)[4]. Considering x_1 as the primary product, the other products $x_2, \ldots, x_n$ may be regarded as "*shadow products*", by analogy with the case of shadow factors[5].

---

[1] Cp. the analogous result in the single-product case (Ch. IV, F above).

[2] Cp. the above case of alternative production with continuous substitution, where the total amount of each input v_i can be allocated to the separate inputs $(v_i = \sum_j v_{ij})$.

[3] Models of this type have been treated by ALLEN (1938), p. 350 (Example 39), and FRISCH (1953), pp. 39 ff.

[4] Cf. Ch. V, B above.

[5] Cf. Ch. V, A above. In the general case (30) the position of the x_j in the model is analogous to that of product shadows; the case of fixed output proportions—(21) above—represents the special case $f_j = b_j f_1$ $(j = 2, 3, \ldots, n)$ where the b_j are constants, resembling models with factor shadows.

The optimum allocation is determined by the production functions (30) and the conditions for maximum profit,

$$\sum_{j=1}^{n} p_j \frac{\delta f_j}{\delta v_i} - q_i = 0 \qquad (i = 1, 2, \ldots, m), \tag{31}$$

or by maximizing $z = \sum_j p_j x_j - c(x_1, \ldots, x_n)$ where the partial marginal costs $c_j{}'$ can be identified with Lagrange multipliers u_j used in minimizing total cost for given parametric x_j, the least-cost conditions being

$$q_i - \sum_{j=1}^{n} u_j \frac{\delta f_j}{\delta v_i} = 0 \qquad (i = 1, 2, \ldots, m)$$

so that

$$dc = \sum_{i=1}^{m} q_i \, dv_i = \sum_{i=1}^{m} \sum_{j=1}^{n} u_j \frac{\delta f_j}{\delta v_i} \, dv_i = \sum_{j=1}^{n} u_j \, dx_j.$$

6. The General Joint-Production Model

From a formal point of view, most of the above models of joint production can be thought of as special cases of a general model[1]

$$F_k(x_1, x_2, \ldots, x_n, v_1, v_2, \ldots, v_m) = 0 \qquad (k = 1, 2, \ldots, M < m + n) \tag{32}$$

where the number of degrees of freedom, $N = m + n - M$, is crucial to the possibilities of product and factor substitution. The extreme cases are $M = 1$, i.e., a model with continuous product and factor substitution ($m + n - 1$ independent variables), and $M = m + n - 1$ ($N = 1$), the case of joint production with fixed output proportions and limitational inputs.

Intermediate models where $1 < M < m + n - 1$ can be rewritten to give M dependent variables in terms of the $m + n - M$ independent variables (inputs and outputs). For example, a model with factor substitution and fixed output proportions—cf. (21) above—represents a case where $M = n$ and the equations have a special form which makes it possible to eliminate the input variables from all but one equation. A model with shadow products such as (30), also characterized by $M = n$, permits each output to be expressed in terms of the m inputs[2].

More generally, the model can be written in parametric form, each input and output variable being a function of $m + n - M$ independent parameters. The cases which have just been discussed can all be written in this form, using a set of input variables for parameters, but the parametric type of model is the more general in that it also includes the linear discontinuous-substitution models where the activity levels serve as parameters.

---

[1] Cf. HENDERSON (1950), pp. 299 f., and (1953), p. 157; RICOSSA (1955); STACKELBERG (1938), pp. 81 f.; ROY (1950), pp. 50 f.; and FRISCH (1953), pp. 38 ff.

[2] FRISCH (1953), pp. 38 ff., has given an instructive classification of the various possible cases, based on the relative magnitudes of m, n, and M and the specific structure of the equations.

Appendix 1

Constrained Maximization

A. Lagrange's Method of Undetermined Multipliers

(a) The typical problem of economic optimization is that of maximizing (or minimizing) some objective function—for example, total cost or total profit—subject to one or more side conditions in the same variables.

Let $g(x_1, x_2, \ldots, x_n)$ be the function to be maximized subject to m constraints

$$f_h(x_1, x_2, \ldots, x_n) = 0 \qquad (h = 1, 2, \ldots, m) \tag{1}$$

where $m < n$ and where g and the f_h are assumed to be differentiable functions. In principle, we can always reduce a problem of this type to the maximization of a function of $n - m$ independent variables, eliminating m dependent variables from the function g by means of equations (1), and set the partial derivatives equal to zero. This method presents no difficulties in the special case where each of the equations (1) gives a particular variable as an explicit function of $n-m$ independent x_i (the same in all equations)[1]. In the more general case, the process of elimination is less easy to handle and some other method must be resorted to.

(b) A simple procedure, which has the added advantage of treating all variables in a symmetrical manner, is provided by the classical method of undetermined multipliers[2], due to LAGRANGE. This method may be assumed to be well-known and a brief summary of the basic principle will suffice.

Form the Lagrangian expression

$$L = g(x_1, x_2, \ldots, x_n) + \sum_{h=1}^{m} u_h f_h(x_1, x_2, \ldots, x_n)$$

where $u_1, u_2, \ldots, u_m$ are provisionally undetermined constants. Then the problem of maximizing the function g subject to (1) is equivalent to that of finding an unconstrained maximum of L, treating $x_1, x_2, \ldots, x_n$ as independent variables. The *necessary* condition for a *maximum* is that the first-order differential vanishes:

---

[1] For example, if profit $z = px - q_1 v_1 - q_2 v_2 - \ldots$ is to be maximized subject to a production function $x = x(v_1, v_2, \ldots)$, the obvious procedure is to take the inputs $v_1, v_2, \ldots$ as independent variables, substituting the production function in the profit expression, which is then maximized with respect to the input variables.

[2] See e.g. OSGOOD (1925), pp. 180 ff.; ALLEN (1938), pp. 364—367 and 498—500; and SAMUELSON (1947), pp. 362—379.

$$dL = \sum_{i=1}^{n} \frac{\delta L}{\delta x_i}\, dx_i = 0,$$

i.e., that the partial derivatives of the Lagrangian L vanish:

$$\frac{\delta L}{\delta x_i} = \frac{\delta g}{\delta x_i} + \sum_{h=1}^{m} u_h \frac{\delta f_h}{\delta x_i} = 0 \qquad (i = 1, 2, \ldots, n). \tag{2}$$

Under "normal" conditions the n equations (2) together with the m equations (1) determine $x_1, x_2, \ldots, x_n$ and $u_1, u_2, \ldots, u_m$ uniquely. If we are not interested in the values of the multipliers u_h, we can eliminate them so that we are left with n equations which determine the x_i. The latter system of equations is easily shown to be identical with the set of necessary maximum conditions obtained if g is reduced to a function of $n - m$ independent variables; hence the two methods of constrained maximization are equivalent[1]. These methods are particularly well suited for the optimization problems of neoclassical economic theory, where the equilibrium positions are typically characterized by marginal equalities—geometrically interpretable as "tangency" conditions—derived from solutions of the form (2).

(c) The same necessary conditions apply if we are looking for a constrained *minimum*. In concrete problems, available additional information will frequently be sufficient to determine whether a point satisfying (1)–(2) is a maximum or a minimum. In the absence of such information, sufficient conditions in terms of second-order derivatives will have to be formulated.

For an unconstrained *maximum* "in the small" of the Lagrangian expression, the *sufficient* condition is

$$d^2L = \sum_{i=1}^{n} \sum_{j=1}^{n} \frac{\delta^2 L}{\delta x_i \delta x_j}\, dx_i\, dx_j < 0 \tag{3}$$

for all variations in the neighbourhood which satisfy

$$df_h = \sum_{i=1}^{n} \frac{\delta f_h}{\delta x_i}\, dx_i = 0 \qquad (h = 1, 2, \ldots, m).$$

For a local *minimum*, the inequality must be reversed. Thus a quadratic form in $dx_1, dx_2, \ldots, dx_n$ is to be negative (positive) definite subject to a

---

[1] In the simple case of $m = 1$, for a maximum of g subject to $f = 0$ we must have

$$dg = g_1'\, dx_1 + g_2'\, dx_2 + \ldots + g_n'\, dx_n = 0$$

for all variations satisfying

$$df = f_1'\, dx_1 + f_2'\, dx_2 + \ldots + f_n'\, dx_n = 0.$$

Eliminating e.g. dx_1, we have with $x_2, \ldots, x_n$ as independent variables

$$dg = (g_2' - g_1' f_2'/f_1')\, dx_2 + \ldots + (g_n' - g_1' f_n'/f_1')\, dx_n = 0.$$

Alternatively, using a Lagrangian multiplier we get

$$dL = (g_1' + u f_1')\, dx_1 + (g_2' + u f_2')\, dx_2 + \ldots + (g_n' + u f_n')\, dx_n = 0.$$

In either case we are led to the necessary condition

$$g_i'/f_i' = g_1'/f_1' \qquad (i = 2, 3, \ldots, n).$$

number of linear restrictions, a criterion which can be shown to involve certain restrictions upon the signs of the principal minors of the determinant whose elements are $\dfrac{\delta^2 L}{\delta x_i \delta x_j}$ "bordered" with the coefficients of the linear restrictions, i.e., $\dfrac{\delta f_h}{\delta x_i}$.

For example, in the case of a single side condition $f(x_1, x_2, \ldots, x_n) = 0$ we have, treating u as a constant and all variables as independent,

$$\frac{\delta^2 L}{\delta x_i \delta x_j} = \frac{\delta}{\delta x_j}\left(\frac{\delta L}{\delta x_i}\right) = \frac{\delta}{\delta x_j}(g_i' + u f_i') = g_{ij}'' + u f_{ij}''$$

from which

$$d^2 L = \sum_{i=1}^{n} \sum_{j=1}^{n} (g_{ij}'' + u f_{ij}'') dx_i dx_j.$$

This quadratic form is negative definite subject to the linear restriction

$$df = \sum_{i=1}^{n} f_i' dx_i = 0$$

if and only if

$$(-1)^k \begin{vmatrix} 0 & f_1' & \cdots & f_k' \\ f_1' & g_{11}'' + u f_{11}'' & \cdots & g_{1k}'' + u f_{1k}'' \\ \cdots & \cdots & & \\ f_k' & g_{k1}'' + u f_{k1}'' & \cdots & g_{kk}'' + u f_{kk}'' \end{vmatrix} > 0 \quad (k = 2, 3, \ldots, n). \quad (4)$$

For a *minimum*, where the quadratic form is to be positive definite, the determinants in (4) must all be negative[1].

The sufficient condition (3), like the first-order conditions (2), can be shown to be equivalent to that derived from the maximization of g as a function of $n - m$ independent variables,

$$d^2 g < 0$$

where

$$df_h = 0 \text{ and } d^2 f_h = 0 \quad (h = 1, 2, \ldots, m)[2].$$

---

[1] Cf. ALLEN (1938), pp. 491 f. For a complete proof of the second-order conditions see BURGER (1955).

[2] In the case of only one side condition $f = 0$, let x_1 be the dependent variable. Differentiating

$$dg = g_1' dx_1 + g_2' dx_2 + \ldots + g_n' dx_n$$

where $dx_2, \ldots, dx_n$ (but not dx_1) are to be treated as constant, we have

$$d^2 g = g_1' d^2 x_1 + dx_1(g_{11}'' dx_1 + \ldots + g_{1n}'' dx_n) + \ldots$$
$$\ldots + dx_n(g_{n1}'' dx_1 + \ldots + g_{nn}'' dx_n)$$

where $df = 0$ and thus also

$$d^2 f = f_1' d^2 x_1 + dx_1(f_{11}'' dx_1 + \ldots + f_{1n}'' dx_n) + \ldots$$
$$\ldots + dx_n(f_{n1}'' dx_1 + \ldots + f_{nn}'' dx_n) = 0.$$

(d) The Lagrangian multipliers have an interesting interpretation relevant to many economic optimization models[1]. Let the side conditions be of the form

$$F_h(x_1, x_2, \ldots, x_n) = b_h \qquad (h = 1, 2, \ldots, m)$$

where the b_h are constants, i.e.,

$$f_h = b_h - F_h(x_1, x_2, \ldots, x_n) = 0.$$

The point at which the function g has a maximum or a minimum subject to the side conditions is determined by (1)–(2). Now, what will be the effect on g of an infinitesimal change in one of the constant terms, say, b_1?

By (2) we have

$$dg = \sum_{i=1}^{n} g_i' \, dx_i = \sum_{i=1}^{n} \sum_{h=1}^{m} u_h \frac{\delta F_h}{\delta x_i} \, dx_i$$

where the variations are to satisfy (1), i.e.,

$$\sum_{i=1}^{n} \frac{\delta F_1}{\delta x_i} \, dx_i = db_1 \text{ and } \sum_{i=1}^{n} \frac{\delta F_h}{\delta x_i} \, dx_i = 0 \qquad (h = 2, 3, \ldots, m).$$

Substituting, we get

$$dg = u_1 \, db_1$$

which means that the marginal change in g with respect to a partial change in one of the constant terms is equal to the Lagrangian multiplier associated with the corresponding side condition[2]. This result holds for any variation from a point satisfying (1)–(2), not only for a movement towards a new maximum (or minimum) position.

B. Linear Programming and the Simplex Criterion

(a) The applicability of Lagrange's method (or the equivalent method of elimination) is subject to three limitations, each of which is serious in some economic optimization models where the optimum point is not a "tangency"

Eliminating d^2x_1 we get (cf. ALLEN (1938), pp. 498 f. and 466)

$$d^2g = \sum_{i=1}^{n} \sum_{j=1}^{n} (g_{ij}'' - f_{ij}'' g_1'/f_1') \, dx_i \, dx_j$$

which is equal to d^2L since, by the necessary conditions,

$$u = -g_1'/f_1'.$$

Substituting this in (4), we get the necessary and sufficient conditions for d^2g to be negative definite subject to $df = 0$, i.e., the sufficient conditions for a (local) maximum of g subject to the side condition $f = 0$.

[1] Cf. SAMUELSON (1947), p. 132, and JOHANSEN (1962).

[2] An important example with a single side condition is cost minimization for constant value of a production function, where the Lagrangian multiplier can be interpreted as marginal cost for small arbitrary variations from the least-cost point. Another example is profit maximization subject to a capacity limitation (in the form of an equality); in this case, u represents the fixed factor's marginal contribution to profit, i.e., its shadow price.

solution characterized by marginal equalities of the type (2) but a "boundary" or "corner" solution.

(*i*) In the first place, the method does not take account of the fact that economic variables are generally defined over the non-negative region. In other words, in addition to the side conditions $f_h = 0$ we also have *non-negativity requirements* $x_i \geq 0$. There is nothing in the method to guarantee that these requirements are automatically satisfied. If a particular point as determined by (1)–(2) happens to respect them, all is well; if it does not, the maximum solution is not feasible, being devoid of economic meaning, and some other method has to be resorted to.

(*ii*) Second, if the functions g and f_h are all *linear*, the partial derivatives $\delta g / \delta x_i$ and $\delta f_h / \delta x_i$ are constant and the necessary conditions cannot be satisfied. This reflects the fact that a linear function (in this case, g as reduced to a function of $n - m$ independent variables) has no finite maximum or minimum[1]—unless the variables are further constrained, for example, by non-negativity requirements, in which case we also run into the difficulty mentioned above.

(*iii*) In the third place, some or all of the side conditions in an economic optimization model may have the form of structural *inequalities*[2] rather than equations, so that we have

$$f_h(x_1, x_2, \ldots, x_n) \geq 0.$$

As with non-negativity requirements, we can always try the Lagrangian method, provisionally disregarding the structural inequalities, and hope for the best. But if it turns out that the solution does not respect them, the solution is not feasible and the method breaks down. Alternatively, we may resort to the device of transforming an inequality into an equation by introducing a "slack" variable $x_h' = f_h$ so that the inequality can be written

$$\varphi_h(x_1, x_2, \ldots, x_n, x_h') = f_h(x_1, x_2, \ldots, x_n) - x_h' = 0.$$

However, the side condition in this form imposes no restriction on the variables $x_1, x_2, \ldots, x_n$, being a mere definition of the additional variable, unless combined with the non-negativity requirement

$$x_h' \geq 0$$

---

[1] As an example, take $g = c_1 x_1 + c_2 x_2$ subject to $a_1 x_1 + a_2 x_2 = b$. Condition (2) becomes
$$c_1 - a_1 u = 0, \quad c_2 - a_2 u = 0$$
which cannot be satisfied simultaneously unless c_1/a_1 happens to be $= c_2/a_2$, in which case we have
$$g = u(a_1 x_1 + a_2 x_2) = u \cdot b = \text{constant}$$
so that there is no question of maximization.

The meaning of this is seen more clearly if we try to solve the problem by elimination of the dependent variable, say, x_1. Solving the side condition for x_1 and substituting in g we have
$$g = c_1 b / a_1 + (c_2 - c_1 a_2 / a_1) x_2$$
which is a linear (i.e., steadily increasing or decreasing) function of the one independent variable x_2 unless the coefficient of x_2 happens to be zero to that g is constant.

[2] Capacity restrictions, for example, will generally be of this form.

to ensure that the inequality $f_h \geq 0$ is not reversed[1]. Again, we do not get very far without a method which can take non-negativity requirements into account.

(b) Such a procedure, applicable to the linear case, has been devised by DANTZIG[2] under the name of the "simplex method".

Suppose we have the problem of maximizing a linear function

$$g = c_1 x_1 + c_2 x_2 + \ldots + c_n x_n$$

subject to m linear restrictions

$$f_h = a_{h1} x_1 + a_{h2} x_2 + \ldots + a_{hn} x_n - b_h = 0 \qquad (h = 1, 2, \ldots, m) \qquad (5)$$

where the variables are required to be non-negative

$$x_i \geq 0 \qquad (i = 1, 2, \ldots, n). \qquad (6)$$

There is no loss of generality in assuming that the side conditions have the form of equations since linear inequalities can always be turned into linear equations by introducing slack variables as indicated above; in this case some of the x_i must be interpreted as slack variables, x_h' having the coefficient 0 in the function g, 1 (or -1) in the h'th side condition, and 0 in all other side conditions, and the slack variables—like the "structural" variables—are required to be ≥ 0. Likewise, a minimum problem can be written in the same form since $\min g = \max (-g)$. Thus, any problem of *linear programming*—i.e., the maximization or minimization of a linear function subject to linear constraints (equations and/or inequalities) and to non-negativity requirements—can be reduced to a problem of the type above.

Because of the linearity of the model, elimination of variables is a easy matter. Therefore, a procedure that suggests itself is that of expressing g in terms of $n - m$ independent variables, taking due account of (6). Taking $x_1, x_2, \ldots, x_m$ as dependent ("basic") variables, we solve the system of m side conditions (5)[3] to get

$$x_i = \bar{x}_i - \sum_{j=m+1}^{n} x_{ij} x_j \qquad (i = 1, 2, \ldots, m) \qquad (7)$$

where $\bar{x}_i$ and the coefficients x_{ij} are constants depending on the a_{hi} and the b_h. Substituting in g we have

---

[1] For example, the linear inequality

$$a_1 x_1 + a_2 x_2 \leq b$$

can be written as the equation

$$a_1 x_1 + a_2 x_2 + x' = b$$

where $x' = b - a_1 x_1 - a_2 x_2 \geq 0$. Had the inequality been

$$a_1 x_1 + a_2 x_2 \geq b,$$

the slack variable would have had the coefficient -1 in the equation.

[2] Cf. DANTZIG (1951).

[3] Assuming that the familiar conditions for a unique solution to exist are satisfied.

$$g = \bar{g} - \sum_{j=m+1}^{n} x_{0j} x_j, \tag{8}$$

where

$$\bar{g} = \sum_{i=1}^{m} c_i \bar{x}_i$$

and

$$x_{0j} = \sum_{i=1}^{m} c_i x_{ij} - c_j$$

are constants depending upon the a_{hi}, b_h, and c_i. Having thus expressed g as a linear function of the independent ("non-basic") variables, how are the x_j to be determined for maximum g?

The clue to this is provided by the fundamental theorem of linear programming, which says that, if the maximum problem can be solved, there exists a maximum solution in which at most m of the n variables are positive, i.e., at least $n - m$ variables are zero[1]. One such "basic" solution is found by putting the $n - m$ independent variables x_j equal to zero in (7)–(8):

$$x_i = \bar{x}_i, \quad x_j = 0, \quad g = \bar{g} \qquad (i = 1, 2, \ldots, m; \ j = m+1, \ldots, n). \tag{9}$$

Let us assume that all $\bar{x}_i \geq 0$. (If this were not so, we would have to pick another basis of m dependent variables which satisfies this condition, and renumber the variables accordingly.) Then our basic solution is *feasible*: it satisfies the side conditions (of which (7) is merely an alternative form) and all of the n variables are non-negative. Whether or not it is also a *maximum* solution, depends on the signs of the coefficients $-x_{0j}$ in (8), the "*simplex coefficients*". If the x_{0j} are all positive, i.e.,

$$\frac{\delta g}{\delta x_j} = -x_{0j} < 0 \qquad (j = m+1, \ldots, n),$$

g is a decreasing function of the x_j. This means that, negative values of the x_j not being admissible, (7)–(8) gives a maximum solution for all $x_j = 0$: any positive x_j will give a smaller value of g. If one of the x_{0j} is $= 0$, the corresponding x_j may be given a positive value (so long as this does not violate the requirement $x_i \geq 0$, cfr. (7)) without the value of g being affected—i.e., the solution is not unique—but (9) is still an optimal solution. In other words,

$$\frac{\delta g}{\delta x_j} = -x_{0j} \leq 0 \qquad (j = m+1, \ldots, n) \tag{10}$$

—a condition known as the "*simplex criterion*"—is a *sufficient* condition for the basic feasible solution (9) to represent a *maximum*[2]. (For a *minimum*,

---

[1] For a simple proof (due to DAVID GALE) see, for example, DANØ (1960), pp. 104 f. For a geometric interpretation see *op. cit.*, pp. 5—9.

[2] Cf. the non-linear case without non-negativity requirements, where the maximum position is characterized by $\delta g / \delta x_j$ being *equal* to zero when g has been expressed in terms of the independent variables x_j, see A above.

the inequality must be reversed.) It is also a *necessary* condition provided that the basic variables are all strictly positive, $\bar{x}_i > 0$. Suppose that (9) is an optimal solution, positive in all x_i, and that the criterion (10) is not satisfied, one of the x_{0j} being negative. Then the solution can be improved by giving the corresponding x_j a small positive value, which is a contradiction. However, if there are zeroes among the $\bar{x}_i$ *("degeneracy")*, this reasoning breaks down because it may not be possible to increase x_j except at the expense of an x_i which is zero in the basic solution[1].

If the solution (9) does not satisfy the simplex criterion, it does not represent a maximum (assuming all $\bar{x}_i > 0$) and another solution must be looked for. It is possible that a maximum solution in more than m positive variables exists, but the fundamental theorem tells us that it cannot yield a larger value of g than the best basic solution. Hence, in principle, the problem can always be solved by testing each of the $\binom{n}{m}$ basic solutions, first for feasibility and next for optimality, using the simplex criterion. However, $\binom{n}{m}$ is a large number when m and n are not very small and the procedure soon becomes impracticable. A short cut by which the computational labour can be reduced very considerably is the iterative procedure known as Dantzig's *simplex algorithm*. Using an arbitrary basic feasible solution as a starting point, the procedure is to improve the solution by moving to a neighbouring basis—i.e., replacing one of the basic variables by an x_j whose x_{0j} is negative—and so forth, until a solution is attained which satisfies the simplex criterion. If degeneracy does not occur, the procedure will always converge and a maximum solution is usually found in a few steps[2]. With a slight modification (due to CHARNES), the simplex method can also be used to deal with cases of degeneracy[3].

In contrast to the Lagrange method (or the equivalent method of eliminating dependent variables), the simplex method is essentially a *numerical* procedure: while it provides a criterion by which a feasible solution can be tested for optimality when the values of the constants are known, as well as an iterative numerical procedure for determining an optimal solution, it cannot yield an *analytical* solution for the variables and for g in terms of the constants a_{hi}, b_h, and c_i.

The optimum solution to a linear programming problem is a *"corner"* solution, geometrically represented by a corner point (in the case of multiple optima, several such points as well as all intermediate points) on the boundary of the feasible region as defined by (5)–(6) and characterized by equilibrium conditions in the form of marginal *inequalities* (the simplex criterion).

(c) In a linear economic model, it will often be of interest to examine in what way the optimum position is affected by changes in the parameters

---

[1] By (7), if $\bar{x}_i = 0$ and $\delta x_i / \delta x_j = -x_{ij} < 0$ for some i, then the x_j in question cannot be increased from zero to a positive value (all other x_j still $= 0$) without making x_i negative, thus violating (6).

[2] For a simple exposition see DANØ (1960), pp. 11—14, 65—74, and 107—109.

[3] See, for example, *op. cit.*, pp. 80—84.

a_{hi}, b_h, or c_i ("parametric linear programming"). Suppose, for example, that one of the right-hand terms in the side conditions (5)—say, b_1—undergoes an infinitesimal change. What will be the effect on g?

Let us assume that the side conditions subject to which we are to maximize $g = \sum\limits_{i=1}^{n} c_i x_i$ have the form of inequalities

$$\sum_{i=1}^{n} a_{hi} x_i \leq b_h \qquad (h = 1, 2, \ldots, m)$$

which can be transformed into equations by introducing non-negative slack variables x_h' $(= x_{n+h})$ to get

$$\sum_{i=1}^{n} a_{hi} x_i + x_h' = b_h,$$

which have the form (5) except that there are now $n + m$ variables.

Now, starting from an optimal basic solution, let b_1 and the corresponding slack variable x_1' be increased by the same infinitesimal amount, $db_1 = dx_1'$. Then the side conditions will be satisfied by a new solution which differs from the original one only in that x_1' is greater than before; the value of g is not affected since the coefficient of x_1' in g is zero. The new solution is feasible, but not necessarily basic and optimal.

Suppose, first, that x_1' is a basic variable, being positive in the original solution. Then, solving for the basic variables, we get for x_1':

$$x_1' = b_1 - \text{(linear terms in the non-basic variables)}$$

and this is the only place where b_1 occurs in the equations corresponding to (7)–(8). It follows that the new solution obtained by adding the same amount to b_1 and x_1' is basic, feasible, and—since the x_{0j} are the same as before—optimal, and the maximum value of g remains the same. Hence the marginal effect on g of a small change in b_1 is zero,

$$dg = 0 \cdot db_1 = 0.$$

If, on the other hand, x_1' is a non-basic variable (i.e., $= 0$) in the original maximum position, the new solution is still feasible but no longer basic since, with $x_1' > 0$, there are now $m + 1$ positive variables (barring degeneracy). Now, by assumption, $\delta g / \delta x_1' \leq 0$. If it is strictly negative, this means that a higher value of g can be attained by reducing x_1' to zero. The position thus reached is a basic, feasible, and optimal solution; the basis and the x_{0j} are the same as in the original position, only the values of the basic variables are different and g has risen by the amount $dg = -(\delta g / \delta x_1') db_1$. This also holds for $\delta g / \delta x_1' = 0$, in which case we have $dg = 0$.

In other words, the effect on g of a marginal change in the constant term b_h is numerically equal to the (negative) simplex coefficient $-x_{0j}$ associated with the corresponding slack variable x_h'. This result is seen to hold whether x_h' is a basic variable or not, if we write (8) (with n non-basic variables) in the form

$$g = \bar{g} - \sum_{i=1}^{n+m} x_{0i} x_i$$

where the basic variables enter with zero coefficients. Then we have for a marginal change in b_h:

$$dg = -(\delta g / \delta x_h') db_h$$

where, by the simplex criterion,

$$-\delta g / \delta x_h' = x_{0, n+h} \begin{cases} = 0 & \text{for } x_h' \text{ basic} \\ \geq 0 & \text{for } x_h' \text{ non-basic.} \end{cases}$$

For purposes of interpretation, it is interesting to note that these coefficients can be identified with the optimal values of the "dual" problem of *minimizing* the function

$$\sum_{h=1}^{m} b_h y_h$$

subject to $y_h \geq 0$ $(h = 1, 2, \ldots, m)$ and to

$$\sum_{h=1}^{m} a_{hi} y_i \geq c_i \qquad (i = 1, 2, \ldots, n)$$

or, with non-negative slack variables y_i',

$$\sum_{h=1}^{m} a_{hi} y_i - y_i' = c_i.$$

By the duality theorem of linear programming[1], in the optimal basic solutions of the two problems we have

$$\sum_{i=1}^{n} c_i x_i = \sum_{h=1}^{m} b_h y_h$$

and

$$y_h = x_{0, n+h},$$

i.e., for a change in b_h we have

$$dg = y_h db_h. \text{[2]}$$

Thus the variables y_h have an interpretation analogous to that of the Lagrange multipliers u_h above.

---

[1] For a proof see DANØ (1960), pp. 109—113. For economic interpretation see *op. cit.*, pp. 92—95.

[2] This result has an important economic interpretation. If the side conditions subject to which profit is to be maximized represent capacity limitations, y_h can be interpreted as the "shadow price" of capacity factor no. h since it represents the increase in profit associated with a unit increase in the capacity b_h.

C. Non-Linear Programming and the Kuhn-Tucker Conditions

(a) The obvious need for generalization of linear programming methods has led to the development of procedures for dealing with non-linear cases[1]. The general problem of *non-linear programming* is that of maximizing (or minimizing) some function $g(x_1, x_2, \ldots, x_n)$ subject to constraints in the form of inequalities

$$f_h(x_1, x_2, \ldots, x_n) \geq 0 \qquad (h = 1, 2, \ldots, m) \tag{11}$$

and to non-negativity constraints

$$x_i \geq 0 \qquad (i = 1, 2, \ldots, n). \tag{12}$$

Linear programming is the special case in which g and all f_h are linear functions.

One way of approaching a problem of this type is to find, as a starting point, an unconstrained maximum of g. If the solution turns out to satisfy (11)–(12), it is an optimal solution to the problem. If, on the other hand, some of the constraints are violated, a new "solution" is computed by maximizing g subject to one of these constraints in equational form (i. e., subject to $f_h = 0$ and/or $x_i = 0$ for the h and i in question). The process is then repeated, adding further constraints one at a time, until a feasible maximum solution emerges [2] [3].

(b) Another approach, due to KUHN and TUCKER, is to adapt the Lagrange method of undetermined multipliers to deal with inequalities[4]. Write the Lagrangian function

$$L = g(x_1, x_2, \ldots, x_n) + \sum_{h=1}^{m} u_h f_h(x_1, x_2, \ldots, x_n).$$

Then the problem of maximizing g subject to (11)–(12) can be shown to be equivalent to that of determining a *saddle point* for L—i.e., a point which represents a maximum with respect to the x_i and a minimum with respect to the u_h—subject to $x_i \geq 0$ and $u_h \geq 0$.

For any differentiable function $L(x_1, \ldots, x_n, u_1, \ldots, u_m)$ where the x_i and the u_h are not confined to the non-negative region, a saddle point is characterized by

$$\frac{\delta L}{\delta x_i} = 0, \quad \frac{\delta L}{\delta u_h} = 0.$$

---

[1] For a survey of non-linear programming methods see DORN (1963).

[2] In certain simple cases of economic optimization this approach suggests itself. For example, if the minimization of the total cost of variable inputs subject to given output leads to a negative value of v_1, the obvious thing to do is to put v_1 equal to zero in the side condition (the production function) and maximize over again. Similarly, if the point of maximum profit violates a capacity restriction $v_1 \leq \bar{v}_1$, the solution will have to be recomputed with $v_1 = \bar{v}_1$.

[3] A numerical method based on this principle has been constructed by THEIL and VAN DE PANNE (1960) to deal with *quadratic programming* problems, where g is a quadratic function whereas the f_h are linear functions. Under certain general conditions the procedure can be shown to lead to an optimal solution in a finite number of steps.

[4] Cf. KUHN and TUCKER (1951).

When it is required that $x_i \geq 0$, $u_h \geq 0$, these necessary conditions must be modified. If some x_i happens to be zero at the saddle point, $\delta L / \delta x_i$ may be

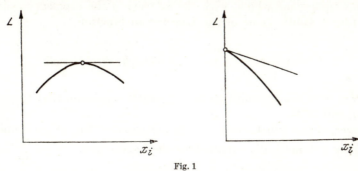

Fig. 1

negative instead of zero, as illustrated in *Fig. 1*. Similarly, $\delta L / \delta u_h$ must be non-negative. Hence the necessary conditions are

$$\frac{\delta L}{\delta x_i} \begin{cases} = 0 \text{ for } x_i > 0 \\ \leq 0 \text{ for } x_i = 0 \end{cases} \tag{13}$$

$$\frac{\delta L}{\delta u_h} \begin{cases} = 0 \text{ for } u_h > 0 \\ \geq 0 \text{ for } u_h = 0, \end{cases} \tag{14}$$

which may be written in the form

$$\frac{\delta L}{\delta x_i} \leq 0, \quad x_i \geq 0, \quad x_i \frac{\delta L}{\delta x_i} = 0 \qquad (i = 1, 2, \ldots, n) \tag{15}$$

$$\frac{\delta L}{\delta u_h} \geq 0, \quad u_h \geq 0, \quad u_h \frac{\delta L}{\delta u_h} = 0 \qquad (h = 1, 2, \ldots, m). \tag{16}$$

Sufficient conditions for a point $(x_1^0, x_2^0, \ldots, x_n^0, u_1^0, u_2^0, \ldots, u_m^0)$ satisfying (15)–(16) to be a saddle point are

$$L(x_1, \ldots, x_n, u_1^0, \ldots, u_m^0) \leq L(x_1^0, \ldots, x_n^0, u_1^0, \ldots, u_m^0)$$
$$+ \sum_{i=1}^{n} \left(\frac{\delta L}{\delta x_i}\right)^0 (x_i - x_i^0) \tag{17}$$

$$L(x_1^0, \ldots, x_n^0, u_1, \ldots, u_m) \geq L(x_1^0, \ldots, x_n^0, u_1^0, \ldots, u_m^0)$$
$$+ \sum_{h=1}^{m} \left(\frac{\delta L}{\delta u_h}\right)^0 (u_h - u_h^0) \tag{18}$$

for all $x_i \geq 0$, $u_h \geq 0$. [1]

Now consider the problem of *maximizing g subject to the constraints* $f_h \geq 0$ *and* $x_i \geq 0$, where g and the f_h are assumed to be differentiable. It can be shown that, subject to certain mild qualifications as to the functions f_h,

---

[1] Proof: Applying (17), (15), (18), and (16) in this order (with $x_i = x_i^0$ and $u_h = u_h^0$ in (15)–(16)), we are led to

$$L(x_1, \ldots, x_n, u_1^0, \ldots, u_m^0) \leq L(x_1^0, \ldots, x_n^0, u_1^0, \ldots, u_m^0) \leq L(x_1^0, \ldots, x_n^0, u_1, \ldots, u_m)$$

for all $x_i \geq 0$ and $u_h \geq 0$, which is the definition of a saddle point. Cf. *op. cit.*, p. 483.

it is *necessary* in order for a point $(x_1{}^0, x_2{}^0, \ldots, x_n{}^0)$ to be a solution of this constrained maximum problem that there exist a set of non-negative Lagrange multipliers $u_h{}^0$ which, with the $x_i{}^0$, satisfy the necessary conditions (15)–(16) for a saddle value of the Lagrangian function

$$L = g(x_1, x_2, \ldots, x_n) + \sum_{h=1}^{m} u_h f_h(x_1, x_2, \ldots, x_n).$$

It is always a *sufficient* condition for a (global) maximum that they satisfy (15), (16), and (17)[1].

Moreover, if the functions g and f_h are *concave*[2] for $x_i \geq 0$, conditions (17)–(18) are automatically satisfied for the Lagrangian $L = g + \sum_h u_h f_h$, so that in this case (15)–(16) are both *necessary* and *sufficient* for a constrained maximum of g as well as for a saddle value of L.[3] These conditions—known as the *Kuhn-Tucker conditions*—can be written

$$\left(\frac{\delta L}{\delta x_i} = \right) \frac{\delta g}{\delta x_i} + \sum_{h=1}^{m} u_h \frac{\delta f_h}{\delta x_i} \leq 0$$

$$x_i \geq 0 \qquad \text{(cf. (12))} \tag{19}$$

$$x_i \left(\frac{\delta g}{\delta x_i} + \sum_{h=1}^{m} u_h \frac{\delta f_h}{\delta x_i} \right) = 0 \qquad (i = 1, 2, \ldots, n)$$

and

$$\left(\frac{\delta L}{\delta u_h} = \right) f_h \geq 0 \qquad \text{(cf. (11))}$$

$$u_h \geq 0 \tag{20}$$

$$u_h f_h = 0 \qquad (h = 1, 2, \ldots, m).$$

---

[1] For proofs, see *op. cit.*, pp. 483—485.

[2] A function $F(x)$ is said to be concave if linear interpolation between any two points never overestimates the actual value of the function at the point of interpolation; cf. the one-variable case illustrated below, where for all x and x^0 in the region of definition we have

$$(1 - \vartheta) F(x^0) + \vartheta F(x) \leq F\{(1 - \vartheta)x^0 + \vartheta x\} \qquad (0 \leq \vartheta \leq 1).$$

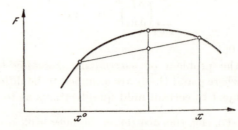

Convexity is defined as the reverse (linear interpolation never underestimates the value of the function). Cf. *op. cit.*, p. 485.

[3] See *op. cit.*, pp. 485 f. The proof follows directly from the definition of concavity.

For a constrained *minimum* of g subject to (11)–(12), the Kuhn-Tucker conditions must be applied to the Lagrangian $L = -g + \sum_{h=1}^{m} u_h f_h$ since a minimum of g is equivalent to a maximum of $(-g)$. Hence, in (19)–(20) $\delta g/\delta x_i$ must be replaced by $-\delta g/\delta x_i$. The necessary conditions are also sufficient for a global minimum if the function g is *convex* (but the f_h concave) because this implies that $(-g)$ is concave.

Conditions (19)–(20) give a criterion which must be satisfied by an optimal solution but do not immediately provide an algorithm to obtain numerical solutions of concrete problems. For certain classes of non-linear programming problems—for example, quadratic programming, where g is a quadratic function and the f_h are linear—computational techniques have been developed to find solutions which satisfy the Kuhn-Tucker conditions[1].

(c) In the case of some or all of the side conditions being *equations* rather than inequalities, the maximum problem can be written in the above form. For example, if the side conditions are

$$f_1 \geq 0, \quad f_2 = 0$$

the side equation is equivalent to the pair of inequalities

$$f_2 \geq 0, \quad f_2 \leq 0$$

and the Lagrangian, with multipliers u, v, w, becomes

$$L = g + u f_1 + v f_2 + w(-f_2).$$

Applying (19)–(20), it is seen that, whereas v and w (as well as u) must be non-negative at the optimal point, no such constraint is placed on their difference $v - w$, the coefficient of f_2 in L. Hence, writing the Lagrangian in the form

$$L = u_1 f_1 + u_2 f_2$$

where $u_1 = u$, $u_2 = v - w$, the Kuhn-Tucker conditions apply with the modification that $\delta L/\delta u_2 = f_2 = 0$ instead of ≥ 0 (the second side condition now being an equation) and that the corresponding multiplier u_2 is no longer required to be non-negative. This result holds for any case of "mixed" side conditions, including the case of all side conditions being equalities.

If $f_h = 0$ for all h and if the requirement $x_i \geq 0$ is dropped, (19)–(20) become the "classical" Lagrange conditions (2) and (1)[2].

Linear programming, where g and the f_h are linear functions, is another special case. The simplex criterion can be shown to be equivalent to the

---

[1] See, e.g., DORN (1963), pp. 180 ff.

[2] Note that in this case the side conditions (1) emerge by differentiating L with respect to the multipliers:

$$\frac{\delta L}{\delta u_h} = f_h = 0$$

whereas we have in the Kuhn-Tucker case

$$\frac{\delta L}{\delta u_h} = f_h \geq 0.$$

Kuhn-Tucker conditions; the sufficiency of (19)–(20) follows from the fact that a linear function is a limiting case of concavity. In the linear case the Lagrange multipliers u_h can be identified with the variables y_h of the dual problem[1], to be interpreted in a similar way as the u_h in the "classical" case. This follows from the symmetry of the Lagrangian with respect to x_i and u_h [2]:

$$L = \sum_{i=1}^{n} c_i x_i + \sum_{h=1}^{m} u_h \left(b_h - \sum_{i=1}^{n} a_{hi} x_i \right)$$

$$= \sum_{h=1}^{m} b_h u_h + \sum_{i=1}^{n} x_i \left(c_i - \sum_{h=1}^{m} a_{hi} u_h \right).$$

Maximizing $\sum_i c_i x_i$ subject to $\sum_i a_{hi} x_i \leq b_h$, $x_i \geq 0$ with multipliers u_h (using the Lagrangian L to find a saddle point (19)–(20)) leads to identically the same conditions as those obtained by minimizing $\sum_h b_h u_h$ subject to $\sum_h a_{hi} u_h \geq c_i$, $u_h \geq 0$ (using the Lagrangian $-L$ with the x_i as multipliers); hence the u_h are the dual variables.

Duality theorems, with dual variables identical with Lagrange multipliers, have also been established for certain classes of non-linear programming problems[3].

---

[1] Cf. above.

[2] Cf. KUHN and TUCKER, *op. cit.*, p. 487.

[3] Cf. DORN (1963), pp. 175 ff.

Appendix 2

Homogeneous Production Functions and the Law
of Variable Proportions

(a) The production curves derived from the production function

$$x = x(v_1,\ v_2)$$

by keeping one of the inputs constant may or may not conform to the
specific law of variable proportions, depending on the shape of the function.
Homogeneity is neither necessary nor sufficient for the law to be valid, but
if the production function is assumed to be homogeneous of the first degree,
it is possible to establish the law on the basis of a relatively small number
of additional simple assumptions.

It will be assumed in the first place that the production function is single-
valued and possesses continuous first- and second-order partial derivatives,
and that a region of substitution exists in which $x_1' > 0$ and $x_2' > 0$ so that

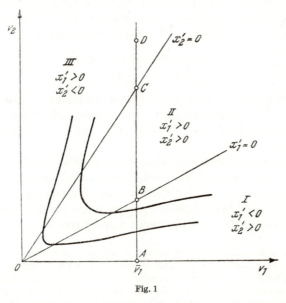

Fig. 1

the isoquants are negatively inclined. Convexity is also assumed in the region. Furthermore, in order for each of the production curves to have a
maximum it must clearly be assumed that the boundaries of the region of

substitution in the factor diagram are positively inclined, dividing the positive factor region into three zones as shown in *Fig. 1*, where Zone II is the region of substitution. Outside this region the isoquants are positively inclined so that Zone I, where $x_1' < 0$, is also characterized by $x_2' > 0$ whereas in Zone III we have $x_1' > 0$, $x_2' < 0$.

Assuming homogeneity of degree one, the marginal productivities are functions of the ratio v_2/v_1 only; hence the boundaries of the region of substitution, $x_1' = 0$ and $x_2' = 0$, will be straight lines passing through the origin. Homogeneity also implies

$$x = v_1 x_1' + v_2 x_2' \tag{1}$$

identically (Euler's theorem); it follows that Zone I is characterized by

$$x_1' < x/v_1, \quad x_2' > x/v_2,$$

Zone II by

$$x_1' < x/v_1, \quad x_2' < x/v_2,$$

and Zone III by

$$x_1' > x/v_1, \quad x_2' < x/v_2.$$

On the boundaries of Zone II we have $x_1' = 0$, $x_2' = x/v_2$ and $x_2' = 0$, $x_1' = x/v_1$ respectively.

Now consider partial variation of, say, v_2 for the other input constant $(v_1 = \bar{v}_1)$, as illustrated in Fig. 1 by the path $ABCD$ which passes through Zones I, II, and III [1]. The corresponding *production curve* (total product) $x = x(\bar{v}_1, v_2)$ will have its *maximum* point at C, where its slope x_2' changes in sign from positive (along AC) to negative (beyond C) through zero. *Average productivity* x/v_2 will have a *maximum* at B, where it is intersected from above by the marginal productivity curve, since

$$\frac{\delta(x/v_2)}{\delta v_2} = \frac{v_2 x_2' - x}{v_2^2}$$

is zero at B, positive in Zone I (AB), and negative in Zones II and III (i.e., beyond B). Correspondingly, the *partial elasticity of production* $\varepsilon_2 = x_2' \cdot v_2/x$ is unity at B, >1 along AB, and <1 beyond B (and also <0 beyond C) [2].

Furthermore, x_2' is decreasing along BC, i.e., $\delta^2 x/\delta v_2^2 = x_{22}'' < 0$, so that the *production curve* $x(\bar{v}_1, v_2)$ is *concave from below between B and C*. This follows from the assumption that the isoquants are convex in Zone II, using also the assumption of homogeneity. At any point on BC, the curvature of the isoquant passing through the point is positive:

---

[1] In examining x as a function of v_2 for $v_1 = \bar{v}_1$ we are not necessarily assuming that v_1 is a capacity factor representing fixed capital equipment. Even if it does, the capacity restriction may have the form $v_1 \leq \bar{v}_1$ rather than a strict equality because the services of the capital equipment are continuously divisible; in this case there is complete symmetry between the two inputs, at least within the capacity limit, so that we might as well have considered—with a similar result—a partial variation in the utilization of the capital equipment, v_1, for given amount of the current input, $v_2 = \bar{v}_2$.

[2] Assuming that $x > 0$ everywhere on the path (cf. below).

$$\frac{d^2v_2}{dv_1^2} = \frac{(x_2')^2 x_{11}'' - 2 x_1' x_2' x_{12}'' + (x_1')^2 x_{22}''}{-(x_2')^3} > 0 \quad (x \text{ constant}).$$

In general nothing is known about the second-order derivatives, but when the production function satisfies Euler's theorem identically we can differentiate (1) partially with respect to v_1 and v_2 to get[1]

$$v_1 x_{11}'' + v_2 x_{21}'' = 0, \quad v_1 x_{12}'' + v_2 x_{22}'' = 0 \tag{2}$$

identically. Using this result to eliminate x_{11}'' and x_{12}'' ($= x_{21}''$) we have

$$\left(\frac{d^2v_2}{dv_1^2} = \right) \frac{(v_2 x_2' + v_1 x_1')^2 \cdot x_{22}''}{-v_1^2 \cdot (x_2')^3} = \frac{x^2 \cdot x_{22}''}{-v_1^2 \cdot (x_2')^3} > 0 \tag{3}$$

in Zone II, where $x_2' > 0$ so that we must have $x_{22}'' < 0$ at any point between B and C. x_{22}'' is also negative at B and C. Since B represents a maximum for x/v_2, we have

$$\frac{\delta^2 (x/v_2)}{\delta v_2^2} = \frac{v_2^3 \cdot x_{22}'' - (v_2 x_2' - x) \cdot 2 v_2}{v_2^4} < 0$$

which, since $v_2 x_2' = x$ at the point, implies $x_{22}'' < 0$ at B. The negativity of x_{22}'' at C, on the other hand, follows directly from the fact that x has a maximum at this point.

On the basis of the assumptions thus far made it is not possible to specify further the shape of the production curve outside the region of substitution. However, if we assume specifically that *output is zero at point A*, i.e., $x(\bar{v}_1, 0) = 0$ (and thus, by the homogeneity of the function, at all points on the axis: $x(v_1, 0) = 0$ for all v_1), we can deduce that *x is positive at any point between A and B*, and that the production curve will have a *point of inflexion* somewhere in this region.

Expanding by Taylor's series with a remainder, starting from an arbitrary point $(\bar{v}_1, h)$ on AB, we have

$$x(\bar{v}_1, 0) = x(\bar{v}_1, h) - h \cdot x_2'(\bar{v}_1, \vartheta h) \text{ for some } \vartheta, \; 0 < \vartheta < 1$$

which, since $x(\bar{v}_1, 0) = 0$, $h > 0$, and $x_2' > 0$, gives $x(\bar{v}_1, h) > 0$ for all h in Zone I. Expanding with an additional term we get

$$x(\bar{v}_1, 0) = x(\bar{v}_1, h) - h \cdot x_2'(\bar{v}_1, h) + \frac{h^2}{2} \cdot x_{22}''(\bar{v}_1, \vartheta h); \tag{4}$$

in Zone I we have $x_2' > x/h$, hence $x_{22}''(\bar{v}_1, \vartheta h) > 0$ for some positive fraction ϑ. For sufficiently small h, $x_{22}''(\bar{v}_1, h)$ will have the same sign, x_{22}'' being continuous by assumption. In other words, x_{22}'' is positive in the neighbourhood of A. Since it is known to be negative in B, there must be at least one point on AB where $x_{22}'' = 0$. Assuming as the simpler case—not unduly restrictive—that there is only one such point of inflexion, P, it represents a *maximum* of the *marginal productivity*. The *production curve*, then, is *convex from below along AP, concave from P to B.*

---

[1] Cf. ALLEN (1938), pp. 317 f.

Beyond C, the production curve is declining ($x_2' < 0$), but *output never gets down to zero* provided that there is symmetry in the sense that output is zero not only on the v_1 axis but on the vertical axis as well; in this case x will be positive at any point in Zone III just as in Zone I. Moreover, there will be a *point of inflexion, Q*, where x_{22}''—negative at C—passes through zero to become positive so that the production curve becomes convex beyond this point. For v_2 constant and v_1 variable there will, by the assumption of symmetry—i.e., $x(0, v_2) = 0$—be some point in Zone III where $x_{11}'' = 0$ and hence also, by (2), $x_{22}'' = 0$. This will hold everywhere on the half-line from 0 through this point since the second-order partial derivatives of a function which is homogeneous of the first degree are homogeneous of degree -1 in v_1 and v_2. The half-line divides Zone III—just as the half-line OP does Zone I—into two regions where x_{11}'' and x_{22}'' are both positive and both negative respectively[1]; the point Q where it intersects $v_1 = \bar{v}_1$ is the second point of inflexion on the curve $x = x(\bar{v}_1, v_2)$.[2]

Thus the production curve for $v_1 = \bar{v}_1$ will have the specific shape of the law of variable proportions, as shown in *Fig. 2*, and so will the curve $x = x(v, \bar{v}_2)$ which we get for v_2 constant[3]. Like the isoquants for different

---

[1] Homogeneity of degree -1 means that we have for any $k > 0$

$$x_{22}''(k v_1, k v_2) = \frac{1}{k} \cdot x_{22}''(v_1, v_2).$$

It follows that x_{22}''—and, by (2), also x_{11}''—will have the same sign—positive, negative, or 0—at any point on the half-line from the origin through (v_1, v_2), the watershed in Zone I being the half-line OP since $x_{22}'' = 0$ at P.

[2] It follows that any isoquant has points of inflexion ($d^2 v_2 / d v_1^2 = 0$, cf. (3)) in Zones I and III where it intersects the half-lines OP and OQ:

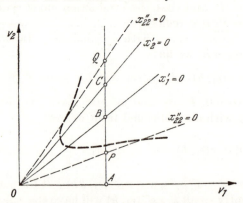

This was not assumed from the outset about the isoquant map of Fig. 1 from which the shape of the production curves was derived; what we have done is to show that it is implicit in the assumption we did make: $x(\bar{v}_1, 0) = 0$ and $x(0, \bar{v}_2) = 0$.

[3] FRISCH (1956), Ch. 6–7, instead of deriving the law of variable proportions from a homogeneous function in two variables, shows that, if the law holds for partial variation of one input and $x = x(v_1, v_2)$ is homogeneous, then it also holds when the other input is varied partially (*op. cit.*, pp. 108 ff.). This comes to much the same thing, as the law is implicit in the assumptions made above. However, FRISCH's production curve differs from that of Fig. 2 in that x becomes $= 0$ somewhere to the right of Q. This, as we have seen, is incompatible

values of x, the production curves obtained for different values of the fixed factor will be radial projections of one another. The most essential characteristics of the production curves—those derived from the assumed existence

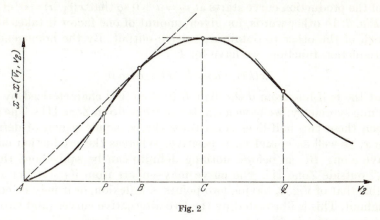

Fig. 2

of the three zones—are summarized in the variation of the partial elasticity ε_2, which runs through the following phases as v_2 is increased for v_1 constant:

Zone I	$\varepsilon_2 > 1$	x/v_2 increasing	
$x_1' = 0$	$\varepsilon_2 = 1$	maximum x/v_2	x increasing
Zone II	$0 < \varepsilon_2 < 1$		
$x_2' = 0$	$\varepsilon_2 = 0$	x/v_2 decreasing	maximum x
Zone III	$\varepsilon_2 < 0$		x decreasing

For v_2 constant, partial variation in v_1 will give a similar picture with respect to ε_1, x, and x/v_1: the path runs through the zones in the reverse order and, by (1), $\varepsilon_1 = 1 - \varepsilon_2$.

It does not necessarily follow that the elasticity decreases monotonically all the way: on the assumptions underlying Fig. 2 all we can say is that

$$\frac{\delta \varepsilon_2}{\delta v_2} = \frac{x \cdot v_2 \cdot x_{22}'' + x_2' \cdot (x - v_2 x_2')}{x^2}$$

is negative along PB and CQ. It seems plausible, though, to assume that ε_2 is also decreasing between B and C, that is, in the region of substitution[1].

Outside the region of substitution (BC), the production curve will be shaped differently if the specific assumption $x(\bar{v}_1, 0) = 0$ is dropped. The curve obviously cannot start on the vertical (x) axis because $x(\bar{v}_1, 0) > 0$ would contradict the basic assumptions underlying the isoquant map of Fig. 1: if there exist Zones I and III where $dv_2/dv_1 > 0$, the isoquants will

with the assumption that $x(\bar{v}_1, 0) = x(0, \bar{v}_2) = 0$, but then FRISCH avoids making such an assumption, letting the curve start at a (small) positive value of the variable input (*op. cit.*, p. 88).

[1] Cf. FRISCH (1956), p. 92.

not have points in common with the v_1 and v_2 axes so that one factor alone cannot produce a positive output. But it is conceivable that output becomes positive only from some point between A and B in Fig. 2, i.e., the positive part of the production curve starts at $v_2 = a > 0$ so that $x(\bar{v}_1, a) = 0$, cf. point R of *Fig. 3*. In other words, for given amount of one factor it takes at least so much of the other to obtain a positive output. By the homogeneity of the production function, we have for $k > 0$

$$x(k\bar{v}_1, ka) = k \cdot x(\bar{v}_1, a) = 0$$

so that the half-line from 0 through R in Zone I is characterized by $x = 0$. Assuming symmetry, we have a similar half-line $0S$ in Zone III[1]. The region between these two half-lines constitutes the economic region of definition where x, as well as v_1 and v_2, is positive. Whereas the production curve is concave along BC as before, nothing definite can be said about the sign of x_{22}'' outside Zone II[2]. The curve may—apart from its starting at R—resemble that of Fig. 2, having two points of inflexion, or it may be concave throughout. This is illustrated by the two alternative curves (and two alternative isoquants) in Fig. 3[3].

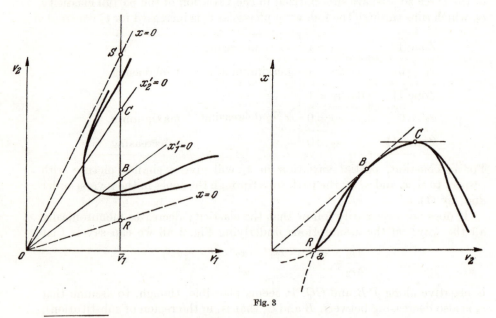

Fig. 3

---

[1] ROS, then, can be said to represent the isoquant $x = 0$.

[2] Corresponding to (4) we can write (for $h > 0$)

$$x(\bar{v}_1, a) = x(\bar{v}_1, a + h) - h \cdot x_2'(\bar{v}_1, a + h) + \frac{h^2}{2} \cdot x_{22}''(\bar{v}_1, a + \vartheta h)$$

for some positive fraction ϑ; we know that $x(\bar{v}_1, a) = 0$, but from the fact that $x_2'(\bar{v}_1, a + h) > x(\bar{v}_1, a + h)/(a + h)$ in Zone I nothing can be deduced as to the sign of $h \cdot x_2' - x$ and thus about the sign of x_{22}'' in the neighbourhood of R.

[3] If the assumption of the three zones is also dropped, assuming instead that x_1' and x_2' are positive throughout the non-negative factor region, the production curve $x = x(\bar{v}_1, v_2)$ will start somewhere on the positive x axis since $x = \bar{v}_1 x_1' + v_2 x_2' > 0$ for $v_2 = 0$ (or, as a

(b) The more general case of a homogeneous production function with any number of inputs can be examined on similar lines for *proportional* variation in some of the inputs, the others being held fast, since this case reduces to a two-factor model. Let the inputs be divided into two groups, the members of which are varied in fixed proportions to v_1 and to v_2 respectively:

$$v_s = \alpha_s \cdot v_1, \quad v_t = \beta_t \cdot v_2.$$

For variations thus restricted, the production function can be written as a homogeneous function of v_1 and v_2,

$$x = x(v_1, v_2, \ldots, v_m) = \Phi(v_1, v_2) = v_1 \Phi_1' + v_2 \Phi_2'$$

where

$$\Phi_1' = x_1' + \sum_s \alpha_s x_s', \quad \Phi_2' = x_2' + \sum_t \beta_t x_t'.$$

Assuming that the straight lines representing $\Phi_1' = 0$ and $\Phi_2' = 0$ in a (v_1, v_2) map are positively inclined and divide the positive quadrant into three zones (as in Fig. 1 above), the production curves

$$x = \Phi(\bar{v}_1, v_2) \text{ and } x = \Phi(v_1, \bar{v}_2)$$

—where v_1 and v_2 now represent homogeneous complexes of inputs measured in units of input no. 1 and 2 respectively—will show diminishing average returns to the variable factor complex from a certain point: for $v_1 = \bar{v}_1$ (and thus $v_s = \alpha_s \bar{v}_1 = $ constant), average productivity x/v_2 will have a maximum at the point where $v_1 = \bar{v}_1$ intersects $\Phi_1' = 0$, and output x will have a maximum for $\Phi_2' = 0$. Correspondingly, the elasticity of production with respect to the variable factor complex, $\varepsilon_2 = (dx/dv_2) \cdot v_2/x = \Phi_2' \cdot v_2/x$—not to be confused with the partial elasticity $x_2' \cdot v_2/x$—will decrease from values greater than 1 through 1 (at the point of maximum x/v_2) and 0 to negative values. For further specification of the shape of the production curve additional assumptions—e.g. $\Phi(\bar{v}_1, 0) = 0$—will have to be introduced[1].

---

limiting case, start at the origin if $x_1' = 0$ coincides with the v_1 axis); the production curve will have no maximum point (x_2' never becomes zero) and there will be diminishing returns everywhere since

$$\frac{\delta(x/v_2)}{\delta v_2} = \frac{v_2 x_2' - x}{v_2^2} = -\frac{\bar{v}_1 x_1'}{v_2^2} < 0.$$

[1] Note, however, that concavity of the production curve in the region of substitution cannot be established unless it is specifically assumed that the isoquants $\Phi(v_1, v_2) = \bar{x}$ are convex in the region. Such convexity does *not* follow from the assumption that the isoquants belonging to the production function $x(v_1, v_2, \ldots, v_m)$ are convex to the origin for $x_i' > 0$ ($i = 1, 2, \ldots, m$). Convexity in m-dimensional factor space means that we have for any two distinct points of the isoquant in the region

(i)
$$\sum_{i=1}^m (x_i')^0 \cdot (v_i - v_i^0) > 0$$

which for variations restricted as above becomes

(ii)
$$(\Phi_1')^0 \cdot (v_1 - v_1^0) + (\Phi_2')^0 \cdot (v_2 - v_2^0) > 0.$$

These inequalities hold for points at which all of the x_i' are positive. Since it is possible for Φ_1' or Φ_2' to be positive even when some of the x_i' are negative, it follows that (ii) does not necessarily hold for all efficient points of the two-dimensional isoquant.

For *unrestricted* variation of some of the inputs of the homogeneous production function, the others being constant, the variation of x is represented by a non-homogeneous production function in terms of the variable inputs only[1]. Geometrical illustration by production curves is no longer possible, but this does not rule out the possibility that some of the features of the law of variable proportions may be revealed as output is expanded along an arbitrary expansion path. Specifically, we can reasonably expect—though it cannot be strictly proved—that the elasticity of production with respect to the (no longer homogeneous) complex of variable factors will take a similar course, passing through the values 1 and 0 [2].

Let the production function $x = x(v_1, v_2, v_3)$ be homogeneous of degree one. On assumptions similar to those made above, variation of v_2 and v_3 in any fixed proportion $\beta = v_3/v_2$ for v_1 constant will conform to the pattern indicated: as output expands, the elasticity $\varepsilon_2 = (dx/dv_2) \cdot v_2/x$ will become $= 1$ at a certain point, later to become zero[3]. Varying the parameter β, the loci of $\varepsilon_2 = 1$ and $\varepsilon_2 = 0$ can be traced out in a (v_2, v_3) map; the latter locus is farther away from the origin. It follows that any expansion path will intersect the two loci in the order mentioned, at least if the path is positively inclined. Recalling that the concept of elasticity of production is not restricted to the special case of a linear path $v_3 = \beta v_2$ with constant β independent of the scale of output—at any point (v_2^0, v_3^0), ε_2 is defined as representing the effect of small proportional factor variations from the point, i.e., for $dv_3 = \beta dv_2$ where $\beta = v_3^0/v_2^0$—the behaviour of ε_2 can be studied along an arbitrary expansion path. For near-proportional variation in v_2 and v_3 the overall variation of ε_2 will be much the same as in the case of proportional variation: expansion along the path will first show increasing returns with respect to the variable factor complex ($\varepsilon_2 > 1$), followed by a phase of decreasing returns ($0 < \varepsilon_2 < 1$) until we get to a point from which output will actually diminish (so that ε_2 becomes negative). If ε_2 decreases monotonically for proportional variation ($v_3 = \beta v_2$)—which is not necessarily the case but

---

[1] This observation is the basis of the familiar tendency to look for hitherto unspecified fixed factors whenever lack of homogeneity manifests itself. The tendency should not be carried too far: while it is always possible to obtain homogeneity by adding dummy factors to the list of inputs—cf. SAMUELSON (1947), p. 84—this has no operational meaning unless the additional factors can be identified with economically relevant input variables which could be varied in fixed proportion to all other inputs with the resulting output actually varying proportionately.

[2] Cf. SCHNEIDER (1934), p. 18.

[3] It does not matter whether the elasticity of production with respect to the complex of variable inputs is taken with respect to v_2 or with respect to v_3: for $v_3 = \beta v_2$ (and v_1 constant) we have

$$\varepsilon_2 = \frac{dx}{dv_2} \cdot \frac{v_2}{x} = \frac{(x_2' + \beta x_3') \cdot v_2}{x} = \frac{v_2 x_2' + v_3 x_3'}{x} = \frac{\left(\frac{1}{\beta}x_2' + x_3'\right) \cdot v_3}{x} = \frac{dx}{dv_3} \cdot \frac{v_3}{x}.$$

The somewhat unsymmetrical notation ε_2 is used here because it refers to a partial variation within the production function $x = \Phi(v_1, v_2)$, cf. above. Alternatively, the symbol ε might have been justified on the ground that it expresses the "total" elasticity of production—the elasticity with respect to scale—when x is considered as a (non-homogeneous) function of the variable inputs, $x = f(v_2, v_3)$. In either case what is meant is the elasticity of $x = x(v_1, v_2, v_3)$ with respect to proportional variation of v_2 and v_3 for v_1 constant.

requires further assumptions—it is likely to do the same along an arbitrary positively inclined path[1]: the family of loci for constant ε_2 will be intersected in the same order by any such path, at least for near-proportional variation[2]. This result is of the greatest importance to the theory of cost; it lends some support to the familiar assumption that marginal cost curves are U-shaped even when there are two or more variable inputs so that the cost function cannot be derived by simple inversion of a production curve. It must be emphasized, however, that the incidence of the law of variable proportions and of the corresponding cost pattern is not confined to cases where the production function is homogeneous of degree one in all input variables.

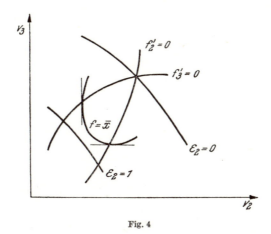

Fig. 4

The region of substitution belonging to the non-homogeneous production function $x = f(v_2, v_3)$—as derived from the homogeneous function $x = x(v_1, v_2, v_3)$ with v_1 constant—will be a closed region, cf. *Fig. 4*. All points at which f_2' $(= x_2'(\bar{v}_1, v_2, v_3))$ and f_3' are both positive will have a positive elasticity of production with respect to proportional variation of the variable factors $(\varepsilon_2 > 0)$, so that the whole of the region of substitution will be "inside" the locus $\varepsilon_2 = 0$ except that the boundaries $f_2' = 0$ and $f_3' = 0$ intersect on the locus, the point of intersection representing the maximum output that can be produced for $v_1 = \bar{v}_1$. [3]

---

[1] The trouble—especially in cases of more than two variable inputs—is that we cannot be sure that the economic expansion path, as defined by parametric cost minimization for given factor prices, will be a monotonically increasing (or at least non-decreasing) curve in factor space. All we can say is that it is likely to be so in "normal" cases.

[2] FRISCH (1956), p. 119, *defines* a "regular" non-homogeneous production function as the case where the elasticity with respect to the variable factor complex decreases monotonically through the values 1 and 0 along any non-decreasing path in factor space, adding (p. 127) that cases of this type can often be ascribed to the constancy of an additional factor which is not specified in the function but which would give a homogeneous function if it had been included in the list of variable inputs. However, it is difficult to specify rigorously on what assumptions homogeneity combined with the constancy of one or more inputs will lead to such regular cases (*op. cit.*, p. 129).

[3] FRISCH, *op. cit.*, p. 124.

References

ALLEN, R. G. D. (1938), Mathematical Analysis for Economists, London: Macmillan.
— (1956), Mathematical Economics, London: Macmillan.
BARFOD, BØRGE (1936), Forenet Produktion og Kvalitetsændring, Nord. Tidsskr. Tek. Økon., lb. nr. 5.
BARGONI, A., B. GIARDINA, and S. RICOSSA (1954), Un esempio di programmazione lineare nell'industria, Ch. 3 in A. BORDIN, ed., La programmazione lineare nell'industria, Torino: Unione Industriale.
BORDIN, ARRIGO (1944), Statica economica, Milano and Messina: Principato.
BREMS, HANS (1952a), A Discontinuous Cost Function, Amer. Econ. Rev., Vol. 42.
— (1957), Input-Output Coefficients as Measures of Product Quality, Amer. Econ. Rev., Vol. 47.
— (1959), Output, Employment, Capital, and Growth, New York: Harper.
— (1951), Product Equilibrium under Monopolistic Competition, Cambridge: Harvard University Press.
— (1952b), En sammenligning mellem den gængse og den Jantzen'ske omkostningsteori, Nationaløkon. Tidsskr., 90. bd.
BURGER, E. (1955), On Extrema with Side Conditions, Econometrica, Vol. 23.
CARLSON, SUNE (1939), A Study on the Pure Theory of Production, London: King.
CHAMBERLIN, EDWARD H. (1933), The Theory of Monopolistic Competition, Cambridge: Harvard University Press.
CHARNES, A., and W. W. COOPER (1961), Management Models and Industrial Applications of Linear Programming, Vol. I–II, New York: Wiley.
CHARNES, A., W. W. COOPER, and DONALD FARR AND STAFF (1953), Linear Programming and Profit Preference Scheduling for a Manufacturing Firm, J. Operations Res. Soc. of America, Vol. 1.
CHENERY, HOLLIS B. (1949), Engineering Bases of Economic Analysis, Cambridge: Harvard University. (Unpublished thesis.)
— (1953), Process and Production Functions from Engineering Data, Ch. 8 in Leontief (1953).
CHIPMAN, JOHN (1953), Linear Programming, Rev. Econ. Statist., Vol. 35.
CHURCHMAN, C. W., R. L. ACKOFF, and E. L. ARNOFF (1957), Introduction to Operations Research, New York: Wiley.
Cost Behavior and Price Policy (1943), New York: National Bureau of Economic Research.
DANØ, SVEN (1960), Linear Programming in Industry: Theory and Applications, Wien: Springer.
— (1965), A Note on Factor Substitution in Industrial Production Processes, Unternehmensforschung, Bd. 9.
DANTZIG, GEORGE B. (1951), Maximization of a Linear Function of Variables Subject to Linear Inequalities, Ch. 21 in KOOPMANS (1951).
DEAN, JOEL (1951), Managerial Economics, New York: Prentice-Hall.
— (1941), The Relation of Cost to Output for a Leather Belt Shop, New York: National Bureau of Economic Research.
DORFMAN, ROBERT (1951), Application of Linear Programming to the Theory of the Firm, Berkeley and Los Angeles: University of California Press.
— (1953), Mathematical, or "Linear", Programming: A Nonmathematical Exposition, Amer. Econ. Rev., Vol. 43.
DORN, W. S. (1963), Non-Linear Programming—A Survey, Management Science, Vol. 9.

Fog, E. S. B. (1934), Blegning af Olier, Kem. Maanedsbl., 15. Aarg.

Forchhammer, N. (1937), Elektriske Glødelampers Levetid og Økonomi, Nord. Tidsskr. Tek. Økon., lb. nr. 8.

Frenckner, T. Paulsson (1957), Betriebswirtschaftslehre und Verfahrensforschung, Z. handelswiss. Forsch., 9. Jahrgang.

Frisch, Ragnar (1932), Einige Punkte einer Preistheorie mit Boden und Arbeit als Produktionsfaktoren, Z. Nat.Ökon., Bd. 3.

— (1956), Innledning til produksjonsteorien, 1. hefte, 8. utg., Oslo: Universitetets Socialøkonomiske Institutt. [English translation: Frisch (1965), Ch. 1–11.]

— (1953), Innledning til produksjonsteorien, 2. hefte, Oslo: Universitetets Socialøkonomiske Institutt. [English translation: Frisch (1965), Ch. 12–19.]

— (1935), The Principle of Substitution. An Example of Its Application in the Chocolate Industry, Nord. Tidsskr. Tek. Økon., lb. nr. 1.

— (1965), Theory of Production, Dordrecht: Reidel.

Georgescu-Roegen, N. (1935), Fixed Coefficients of Production and the Marginal Productivity Theory, Rev. Econ. Stud., Vol. 3.

Gloerfelt-Tarp, B. (1937), Den økonomisk definerede Produktionsfunktion og den heterogene Fremstillingsproces, Nord. Tidsskr. Tek. Økon., lb. nr. 10.

Grosse, Anne P. (1953), The Technological Structure of the Cotton Textile Industry, Ch. 10 in Leontief (1953).

Gutenberg, Erich (1963), Grundlagen der Betriebswirtschaftslehre, Erster Band: Die Produktion, 8./9. Aufl., Berlin: Springer.

Heady, Earl O., and John Pesek (1960), Expansion Paths for Some Production Functions, Econometrica, Vol. 28.

Henderson, John S. (1953), Marginal Productivity Analysis—A Defect and a Remedy, Econometrica, Vol. 21.

— (1950), Production with Limitational Goods, Econometrica, Vol. 18.

Hicks, J. R. (1946), Value and Capital, 2nd ed., Oxford: Clarendon.

Jantzen, Ivar (1939), Basic Principles of Business Economics and National Calculation, Copenhagen: Gad.

— (1924), Voxende Udbytte i Industrien, Nationaløkon. Tidsskr., 62. Bd.

Johansen, Leif (1962), Notat om tolkingen av Lagrange-multiplikatorer, Memorandum fra Socialøkonomisk Institutt, Oslo.

Kaldor, Nicholas (1937), Limitational Factors and the Elasticity of Substitution, Rev. Econ. Stud., Vol. 4.

Knight, Frank H. (1921), Risk, Uncertainty and Profit, Boston: Houghton Mifflin.

Koopmans, Tjalling C. (1951), ed., Activity Analysis of Production and Allocation, Cowles Commission Monograph No. 13, New York: Wiley.

Kuhn, H. W., and A. W. Tucker (1951), Nonlinear Programming, pp. 481–492 in J. Neyman, ed., Proceedings of the Second Berkeley Symposium on Mathematical Statistics and Probability, Berkeley and Los Angeles: University of California Press.

Leontief, Wassily (1941), The Structure of American Economy, 1919–1929, Cambridge: Harvard University Press.

Leontief, Wassily, and others (1953), Studies in the Structure of the American Economy, New York: Oxford University Press.

Osgood, William F. (1925), Advanced Calculus, New York: Macmillan.

Pareto, Vilfredo (1897), Cours d'économie politique, Vol. II, Lausanne: Rouge.

— (1927), Manuel d'économie politique, 2e éd., Paris: Giard.

Pedersen, H. Winding (1949), Omkostninger og prispolitik, 2. udg., København: Høst.

— (1933), Omkring Kapacitetsudnyttelsesteorien, Nationaløkon. Tidsskr., 71. Bd.

Perry, John H. (1941), ed., Chemical Engineers' Handbook, 2nd ed., New York: McGraw-Hill.

Pfouts, Ralph W. (1961), The Theory of Cost and Production in the Multi-Product Firm, Econometrica, Vol. 29.

Ricossa, Sergio (1955), Nuove applicazioni della teoria economica della produzione, G. Econo., Vol. 14.

— (1954), Nuovi e vecchi problemi della produzione, Note Econometriche, N. 11.

RIEBEL, PAUL (1955), Die Kuppelproduktion: Betriebs- und Marktprobleme, Köln: Westdeutscher Verlag.

ROY, RENÉ (1950), Remarques sur les phénomènes de production, Metroeconomica, Vol. 2.

SAMUELSON, PAUL A. (1947), Foundations of Economic Analysis, Cambridge: Harvard University Press.

SCHMIDT, ERIK (1939), Økonomisk definerede produktionsfunktioner, Nord. Tidsskr. Tek. Økon., lb. nr. 17–18.

SCHNEIDER, ERICH (1958), Einführung in die Wirtschaftstheorie, II. Teil, 5. verb. u. erw. Aufl., Tübingen: Mohr.

— (1937), Kapacitetsudnyttelsesproblemets to Dimensioner, Nord. Tidsskr. Tek. Økon., lb. nr. 7.

— (1942), Teoria della produzione, Milano: Casa Editrice Ambrosiana.

— (1934), Theorie der Produktion, Wien: Springer.

— (1940), Über den Einfluß von Leistung und Beschäftigung auf Kosten und Erfolg einer Unternehmung mit homogener Massenfabrikation, Arch. math. Wirtsch., Bd. 6.

SCHULTZ, HENRY (1929), Marginal Productivity and the General Pricing Process, J. Polit. Econ., Vol. 37.

SCHWEYER, HERBERT E. (1955), Process Engineering Economics, New York: McGraw-Hill.

SHEPHARD, RONALD W. (1953), Cost and Production Functions, Princeton: University Press.

STACKELBERG, HEINRICH V. (1938), Angebot und Nachfrage in der Produktionswirtschaft, Arch. math. Wirtsch., Bd. 4.

— (1932), Grundlagen einer reinen Kostentheorie, Wien: Springer.

— (1941), Stundenleistung und Tagesleistung, Arch. math. Wirtsch., Bd. 7.

STIGLER, GEORGE J. (1939), Production and Distribution Theories in the Short Run, J. Polit. Econ., Vol. 47.

— (1946), Production and Distribution Theories: The Formative Period, New York: Macmillan.

SYMONDS, GIFFORD H. (1955), Linear Programming: The Solution of Refinery Problems, New York: Esso Standard Oil Co.

THEIL, H., and C. VAN DE PANNE (1960), Quadratic Programming as an Extension of Classical Quadratic Maximization, Management Science, Vol. 7.

WALRAS, LÉON (1954), Elements of Pure Economics, Translated by WILLIAM JAFFÉ, London: Allen & Unwin.

WALTERS, A. A. (1963), Production and Cost Functions: An Econometric Survey, Econometrica, Vol. 31.

WICKSTEED, PHILIP H. (1894), An Essay on the Co-ordination of the Laws of Distribution, London: Macmillan.

ZEUTHEN, F. (1955), Economic Theory and Method, London: Longmans.

— (1933), Das Prinzip der Knappheit, technische Kombination und ökonomische Qualität, Z. Nat.Ökon., Bd. 4.

— (1928), Den økonomiske Fordeling, København: Busck.

— (1942), Økonomisk Teori og Metode, København: Busck.

Index

Druck R. Spies & Co., Wien